About the Author

Professor Luc Ferry teaches philosophy at the Sorbonne (University of Paris VII, Denis Diderot). He has won numerous awards, including the Prix Médicis, Prix Jean-Jacques-Rousseau, and Prix Aujourd'hui, in addition to being an officer of the French Legion of Honor and a knight of the Order of Arts and Letters. From 2002 to 2004 Ferry served as French Minister of National Education. He lives in Paris.

A BRIEF HISTORY *of* THOUGHT

Also by Luc Ferry, available in English

French Philosophy of the Sixties: An Essay on Antihumanism (with Alain Renaut)

Heidegger and Modernity (with Alain Renaut)

Homo Aestheticus: The Invention of Taste in the Democratic Age

The New Ecological Order

Man Made God: The Meaning of Life

Political Philosophy

What Is the Good Life?

As Editor

Why We Are Not Nietzscheans (with Alain Renaut)

A BRIEF HISTORY *of* THOUGHT

A PHILOSOPHICAL GUIDE TO LIVING

LUC FERRY

Translated by Theo Cuffe

HARPER PERENNIAL

NEW YORK • LONDON • TORONTO • SYDNEY • NEW DELHI • AUCKLAND

HARPER PERENNIAL

First published under the title *Learning to Live* in Great Britain in 2010 by Canongate Books.

HarperCollins books may be purchased for educational, business, or sales promotional use. For information, please e-mail the Special Markets Department at SPsales@harpercollins.com.

FIRST U.S. EDITION

Designed by Palimpset Book Production Limited

Library of Congress Cataloging-in-Publication Data is available upon request.

ISBN 978-0-06-207424-9

16 17 18 19 20 OV/RRD 20 19 18 17 16 15 14 13 12 11

For Gabrielle, Louise and Clara

CONTENTS

INTRODUCTION

While chatting over supper on holiday, some friends asked me to improvise a philosophy course for adults and children alike. I decided to accept the challenge and came to relish it. The exercise forced me to stick to essentials – no complicated words, no learned quotations and no references to obscure theories. As I worked through my account of the history of ideas, without access to a library, it occurred to me that there is nothing comparable in print. There are many histories of philosophy, of course; some are excellent, but even the best ones are a little dry for someone who has left university behind, and certainly for those yet to enter a university. And the rest of us are not particularly concerned.

This book is the direct result of those evenings amongst friends, so I have tried to preserve the original impromptu style. Its objective is both modest and ambitious: modest, because it is addressed to a nonacademic audience; ambitious, because I have not permitted myself any concession to simplification where it would involve distortion of the philosophical ideas at its heart. I feel too much respect for the masterpieces of philosophy to caricature them. Clarity should be the primary responsibility of a work addressed to beginners, but it must be achieved without compromising the truth of its subject; otherwise it is worthless.

With that in mind, I have tried to offer a rite of passage, which aims to be as straightforward as possible, without bypassing the richness and profundity of philosophical ideas. My aim is not merely to give a taste, a superficial gloss, or a survey influenced by popular trends; on the contrary I want to lay bare these ideas in their integrity, in order to satisfy two needs: that of an adult who wants to know what philosophy is about, but does not necessarily intend to proceed any further; and that of a young person who hopes eventually to further their study, but does not as yet have the necessary bearings to be able to read these challenging authors for herself or himself.

I have attempted to give an account of everything that I consider to be truly indispensable in the history of thought – all that I would like to pass on to family and those whom I regard as friends.

But why undertake this endeavour? First, because even the most sublime spectacle begins to pall if one lacks a companion with whom to share it. I am increasingly aware that philosophy no longer counts as what is ordinarily thought of as 'general knowledge'. An educated person is supposed to know his or her national history, a few standard literary and artistic references, even a few odds and ends of biology or physics, yet they most likely have no inkling of Epictetus, Spinoza or Kant. I am convinced that everyone should study just a little philosophy, if only for two simple reasons.

First of all, without it we can make no sense of the world in which we live. Philosophy is the best training for living, better even than history and the human sciences. Why? Quite simply because virtually all of our thoughts, convictions and values exist and have meaning – whether or not we are conscious of it – within models of the world that have been developed over the course of intellectual history. We must understand these models in order to grasp their reach, their logic and their consequences.

Many individuals spend a considerable part of their lives anticipating misfortune and preparing for catastrophe – loss of work, accident, illness, death of loved ones, and so on. Others, on the contrary, appear to live in a state of utter indifference, regarding such fears as morbid and having no place in everyday life. Do they realise, both of these character-types, that their attitudes have already been pondered with matchless profundity by the philosophers of ancient Greece?

The choice of an egalitarian rather than an aristocratic ethos, of a romantic aesthetic rather than a classical one, of an attitude of attachment or non-attachment to things and to beings in the face of death; the adoption of authoritarian or liberal political attitudes; the preference for animals and nature over mankind, for the call of the wild over the cities of man – all of these choices and many more were considered long before they became opinions available, as in a marketplace, to the citizen. These divisions, conflicts and issues continue to determine our thoughts and our words, whether we are aware of them or not. To study them in their pure form, to grasp their deepest origins, is to arm oneself with not only the means of becoming more intelligent, but also more independent. Why would one deprive oneself of such tools?

Second, beyond coming to an understanding of oneself and others through acquaintance with the key texts of philosophy, we come to realise that these texts are able, quite simply, to help us live in a better and freer way. As several contemporary thinkers note: one does not philosophise to amuse oneself, nor even to better understand the world and one's own place in it, but sometimes literally to 'save one's skin'. There is in philosophy the wherewithal to conquer the fears which can paralyse us in life, and it is an error to believe that modern psychology, for example, can substitute for this.

Learning to live; learning to fear no longer the various faces of death; or, more simply, learning to conquer the banality of everyday life – boredom, the sense of time slipping by: these were already the primary motivations of the schools of ancient Greece. Their message deserves to be heard, because, contrary to what happens in history and in the human sciences, the philosophers of time past speak to us in the present tense. And this is worth contemplating.

When a scientific theory is revealed to be false, when it is refuted by another manifestly truer theory, it becomes obsolete and is of no further interest except to a handful of scientists and historians. However, the great philosophical questions about how to live life remain relevant to this day. In this sense, we can compare the history of philosophy to that of art, rather than of the sciences: in the same way that paintings by Braque or Kandinsky are not 'less beautiful' than those by Vermeer or Manet, so too the reflections of Kant or Nietzsche on the sense or non-sense of life are not inferior – or superior – to those of Epictetus, Epicurus or the

Buddha. They all furnish propositions about life, attitudes in the face of existence, that continue to address us across the centuries. Whereas the scientific theories of Ptolemy or Descartes may be regarded as 'quaint' and have no further interest other than the historical, we can still draw upon the collective wisdom of the ancients as we can admire a Greek temple or a Chinese scroll – with both feet planted firmly in the twenty-first century.

Following the lead of the earliest manual of philosophy ever written, *The Discourses* of Epictetus from *c.* 100 AD, this little book will address its readers directly. I hope the reader may take my tone as a sign of complicity rather than familiarity.

A BRIEF HISTORY *of* THOUGHT

1
WHAT IS PHILOSOPHY?

I am going to tell you the story as well as the history of philosophy. Not all of it, of course, but its five great moments. In each case, I will give you an example of one or two transforming visions of the world or, as we say sometimes, one or two great 'systems of thought'. I promise that, if you take the trouble to follow me, you will come to understand this thing called philosophy and you will have the means to investigate it further – for example, by reading in detail some of the great thinkers of whom I shall be speaking.

The question 'What is philosophy?' is unfortunately one of the most controversial (although in a sense that is a good thing, because we are forced to exercise our ability to reason) and one which the majority of philosophers still debate today, without finding common ground.

When I was in my final year at school, my teacher assured me that it referred 'quite simply' to the 'formation of a critical and independent spirit', to a 'method of rigorous thought', to an 'art of reflection', rooted in an attitude of 'astonishment' and 'enquiry' . . . These are the definitions which you still find today in most introductory works. However, in spite of the respect I have for my teacher, I must tell you from the start that, in my view, such definitions have nothing to do with the question.

It is certainly preferable to approach philosophy in a reflective spirit; that much is true. And that one should do so with rigour and even in a critical and interrogatory mood – that is also true. But all of these definitions are entirely non-specific. I'm sure that you can think of an infinite number of other human activities about which we should also ask questions and strive to argue our way as best we can, without their being in the slightest sense philosophical.

Biologists and artists, doctors and novelists, mathematicians and theologians, journalists and even politicians all reflect and ask themselves questions – none of which makes them, for my money, philosophers. One of the principal errors of the contemporary world is to reduce philosophy to a straightforward matter of 'critical reflection'. Reflection and argument are worthy activities; they are indispensable to the formation of good citizens and allow us to participate in civic life with an independent spirit. But these are merely the means to an end – and philosophy is no more an instrument of politics than it is a prop for morality.

I suggest that we accept a different approach to the question 'What is philosophy?' and start from a very simple proposition, one that contains the central question of all philosophy: that the human being, as distinct from God, is mortal or, to speak like the philosophers, is a 'finite being', limited in space and time. As distinct from animals, moreover, a human being is the only creature who is aware of his limits. He knows that he will die, and that his near ones, those he loves, will also die. Consequently he cannot prevent himself from thinking about this state of affairs, which is disturbing and absurd, almost unimaginable. And, naturally enough, he is inclined to turn first of all to those religions which promise 'salvation'.

The Question of Salvation

Think about this word – 'salvation'. I will show how religions have attempted to take charge of the questions it raises. Because the simplest way of starting to define philosophy is always by putting it in relation to religion.

Open any dictionary and you will see that 'salvation' is defined first and foremost as 'the condition of being saved, of escaping a great danger or misfortune'. But from what 'great danger', from what 'misfortune' do religions claim to deliver us? You already know the answer: from the peril of death. Which is why all religions strive, in different ways, to promise us eternal life; to reassure us that one day we will be reunited with our loved ones – parents and friends, brothers and sisters, husbands and wives, children and grandchildren – from whom life on earth must eventually separate us.

In the Gospel According to St John, Jesus experiences the death of a dear friend, Lazarus. Like every other human being since the dawn of time, he weeps. He experiences, like you or I, the grief of separation. But unlike you or I, simple mortals, it is in Jesus's power to raise his friend from the dead. And he does this in order to prove that, as he puts it, 'love is stronger than death'. This fundamental message constitutes the essence of the Christian doctrine of redemption: death, for those who love and have faith in the word of Christ, is but an appearance, a rite of passage. Through love and through faith, we shall gain immortality.

Which is fortunate for us, for what do we truly desire, above all else? To be understood, to be loved, not to be alone, not to be separated from our loved ones – in short, not to die and not to have them die on us. But daily life will sooner or later disappoint every one of these desires, and, so it is, that by trusting in a God some of us seek salvation, and religion assures us that those who do so will be rewarded. And why not, for those who believe and have faith?

But for those who are not convinced, and who doubt the truth of these promises of immortality, the problem of death remains unresolved. Which is where philosophy comes in. Death is not as simple an event as it is ordinarily credited with being. It cannot merely be written off as 'the end of life', as the straightforward termination of our existence. To reassure themselves, certain wise men of antiquity (Epicurus for one) maintained that we must not think about death, because there are only two alternatives: either I am alive, in which case death is by definition elsewhere; or death is here and, likewise by definition, I am not here to worry about it! Why, under these conditions, would you bother yourself with such a pointless problem?

This line of reasoning, in my view, is a little too brutal to be honest. On the contrary, death has many different faces. And it is this which torments man: for only man is aware that his days are numbered, that the inevitable is not an illusion and that he must consider what to do with his brief existence. Edgar Allan Poe, in one of his most famous poems, 'The Raven', conveys this idea of life's irreversibility in a sinister raven perched on a window ledge, capable only of repeating 'Nevermore' over and over again.

Poe is suggesting that death means *everything that is unrepeatable*. Death is, *in the midst of life*, that which will not return; that which belongs irreversibly to time past, which we have no hope of ever recovering. It can mean childhood holidays with friends, the divorce of parents, or the houses or schools we have to leave, or a thousand other examples: even if it does not always mean the disappearance of a loved one, everything that comes under the heading of 'Nevermore' belongs in death's ledger.

In this sense, you can see how far death is from a mere biological ending. We encounter an infinite number of its variations, in the midst of life, and these many faces of death trouble us, even if we are not always aware of them. To live well, therefore, to live freely, capable of joy, generosity and love, we must first and foremost conquer our fear – or, more accurately, our fears of the irreversible. But here, precisely, is where religion and philosophy pull apart.

Philosophy versus Religion

Faced with the supreme threat to existence – death – how does religion work? Essentially, through faith. By insisting that it is faith, and faith alone, which can direct the grace of God towards us. If you believe in Him, God will save you. The religions demand *humility*, above and beyond all other virtues, since humility is in their eyes the opposite – as the greatest Christian thinkers, from Saint Augustine to Pascal, never stop telling us – of the arrogance and the vanity of philosophy. Why is this accusation levelled against free thinking? In a nutshell, because philosophy *also* claims to save us – if not from death itself, then from the anxiety it causes, and to do so *by the exercise of our own resources and our innate faculty of reason.* Which, from a religious perspective, sums up philosophical pride: the effrontery evident already in the earliest philosophers, from Greek antiquity, several centuries before Christ.

Unable to bring himself to believe in a God who offers salvation, the philosopher is above all one who believes that by understanding the world, by understanding ourselves and others as far our intelligence permits, we shall succeed in overcoming fear, through clear-sightedness rather than blind faith.

In other words, if religions can be defined as 'doctrines of salvation', the great philosophies can also be defined as doctrines of salvation (but without the help of a God). Epicurus, for example, defined philosophy as 'medicine for the soul', whose ultimate aim is to make us understand that 'death is not to be feared'. He proposes four principles to remedy all those ills related to the fact that we are mortal: 'The gods are not to be feared; death cannot be felt; the good can be won; what we dread can be conquered.' This wisdom was interpreted by his most eminent disciple, Lucretius, in his poem *De rerum natura* ('On the Nature of Things'):

> The fear of Acheron [the river of the Underworld] must first and foremost be dismantled; this fear muddies the life of man to its deepest depths, stains everything with the blackness of death, leaves no pleasure pure and clear.

And Epictetus, one of the greatest representatives of another of the ancient Greek philosophical schools – Stoicism – went so far as to reduce *all* philosophical questions to a single issue: the fear of death. Listen for a moment to him addressing his disciple in the course of his dialogues or *Discourses*:

> Keep well in mind, then, that this epitome of all human evils, of mean-spiritedness and cowardice, is not death as such, but rather the fear of death. Discipline yourself, therefore, against this. To which purpose let all your reasonings, your readings, all your exercises tend, and you will know that only in this way are human beings set free. (*Discourses,* III, 26, 38–9)

The same theme is encountered in Montaigne's famous adage – 'to philosophise is to learn how to die'; and in Spinoza's reflection about the wise man who 'dies less than the fool'; and in Kant's question, 'What are we permitted to hope for?' These references may mean little to you, because you are only starting out, but we shall come back to each of them in turn. Bear them in mind. All that matters, now, is that we understand why, in the eyes of every philosopher, fear of death prevents us from living – and not only because it generates anxiety. Most of the time, of course, we do not meditate on human mortality. But at a deeper level the irreversibility of things is a kind of death at the heart of life and threatens constantly to steer us into *time past* – the home of nostalgia, guilt, regret and remorse, those great spoilers of happiness.

Perhaps we should try not to think of these things, and try to confine ourselves to happy memories, rather than reflecting on bad times. But paradoxically those happy memories can become transformed, over time, into 'lost paradises', drawing us imperceptibly towards the past and preventing us from enjoying the present.

Greek philosophers looked upon the past and the future as the primary evils weighing upon human life, and as the source of all the anxieties which blight the present. The present moment is the only dimension of existence worth inhabiting, because it is the only one available to us. The past is no longer and the future has yet to come, they liked to remind us; yet we live virtually all of our lives somewhere between memories and aspirations, nostalgia and expectation. We imagine we would be much happier with new shoes, a faster computer, a bigger house, more exotic holidays, different friends . . . But by regretting the past or guessing the future, we end up missing the only life worth living: the one which proceeds from the here and now and deserves to be savoured.

Faced with these mirages which distract us from life, what are the promises of religion? That we don't need to be afraid, because our hopes will be fulfilled. That it is possible to live in the present as it is – *and* expect a better future! That there exists an infinitely benign Being who loves us above all else and will therefore save us from the solitude of ourselves and from the loss of our loved ones, who, after they die in this world, will await us in the next.

What must we do to be 'saved'? Faced with a Supreme Being, we are invited to adopt an attitude framed entirely in two words: trust (Latin *fides*, which also means 'faith') and humility. In contrast, philosophy, by following a different path, verges on the *diabolical*. Christian theology developed a powerful concept of 'the temptations of the devil'. Contrary to the popular imagery which frequently served the purposes of a Church in need of authority, the devil is not one who leads us away from the straight and narrow, morally speaking, by an appeal to the weaknesses of the flesh. The devil is rather one who, spiritually speaking, does everything in his power to separate us (*dia-bolos* in Greek meaning 'the who who divides') from the vertical link uniting true believers with God, and which alone saves them from solitude and death. The *diabolos* is not content with setting men against each other, provoking them to hatred and war, but much more ominously, he cuts man off from God and thus delivers him back into the anguish that faith had succeeded in healing.

For a dogmatic theologian, philosophy is the devil's own work, because by inciting man to turn aside from his faith, to exercise his reason and give rein to his enquiring spirit, philosophy draws him imperceptibly into the realm of *doubt*, which is the first step beyond divine supervision.

In the account of Genesis, with which the Bible opens, the serpent plays the role of Devil by encouraging Adam and Eve – the first human beings – to doubt God's word about the forbidden fruit. The serpent wants them to ask questions and try the apple, so that they will disobey God. By separating them from Him, the Devil can then inflict upon them – mere mortals – all the torments of earthly existence. The 'Fall' of Adam and Eve and their banishment from the first Paradise is the direct consequence of *doubting* divine edicts; thus, men became mortal.

All philosophies, however divergent they may sometimes be in the answers they bring, promise us an escape from primitive fears. They possess in common with religions the conviction that anguish prevents us from leading good lives: it stops us not only from being happy, but also from being free. This is an ever present theme amongst the earliest Greek philosophers: we can neither think nor act freely when we are paralysed by the anxiety provoked – even unconsciously – by fear of the irreversible. The question becomes one of how to persuade humans to 'save' themselves.

Salvation must proceed not from an Other – from some Being supposedly transcendent (meaning 'exterior to and superior to' ourselves) – but well and truly from within. Philosophy wants us to get ourselves out of trouble by utilising our own resources, by means of reason alone, with boldness and assurance. And this of course is what Montaigne meant when, characterising the wisdom of the ancient Greeks, he assured us that 'to philosophise is to learn how to die'.

Is every philosophy linked therefore to atheism? Can there not be a Christian or a Jewish or a Muslim philosophy? And if so, in what sense? In other words, what are we to make of those philosophers, like Descartes or Kant, who believed in God? And you may ask why should we refuse the promise of religion? Why not submit with humility to the requirements of salvation 'in God'?

For two crucial reasons, which lie at the heart of all philosophy. First and foremost, because the promise of religions – that we are immortal and will encounter our loved ones after our own biological demise – is too good to be true. Similarly hard to believe is the image of a God who acts as a father to his children. How can one reconcile this with the appalling massacres and misfortunes which overwhelm humanity: what father would abandon his children to the horror of Auschwitz, or Rwanda, or Cambodia? A believer will doubtless respond that that is the price of freedom, that God created men as equals and evil must be laid at their door. But what about the innocent? What about the countless children martyred in the course of these crimes against humanity? A philosopher begins to doubt that the religious answers are adequate. (Undoubtedly this argument engages only with the popular image of religion, but this is nonetheless the most widespread and influential version available.) Almost invariably the philosopher comes to think that belief in God, which usually arises as an indirect consequence, in the guise of consolation, perhaps makes us lose in clarity what we gain in serenity. He respects all believers, it goes without saying. He does not claim that they are necessarily wrong, that their faith is absurd, or that the non-existence of God is a certainty. (How would one set about proving that God does not exist?) Simply, that in his case there is a failure of faith; therefore he must look elsewhere.

Wellbeing is not the only ideal in life. Freedom is another. And if religion calms anguish by making death into an illusion, it risks doing so at the price of freedom of thought. For it demands, more or less, that we abandon reason and the enquiring spirit in return for faith and serenity. It asks that we conduct ourselves, before God, like little children, not as curious adults.

Ultimately, to philosophise, rather than take on trust, is to prefer lucidity to comfort, freedom rather than faith. It also means, of course, 'saving one's skin', but not at any price. You might ask, if philosophy is essentially a quest for a good life beyond the confines of religion – a search for salvation without God – why is it so frequently presented in books as the art of right-thinking, as the exercise of the critical faculty and freedom of conscience? Why, in civic life, on television and in the press, is philosophy so often reduced to moral engagement, casting the vote for justice and against injustice? The philosopher is portrayed as someone who understands things as they are, who questions the evils of the day. What are we to make of the intellectual and moral life, and how do we reconcile these imperatives with the definition of philosophy I have just outlined?

The Three Dimensions of Philosophy

If the quest for a salvation without God is at the heart of every great philosophical system, and that is its essential and ultimate objective, it cannot be accomplished without deep reflection upon reality, or things as they are – what is ordinarily called 'theory' – and consideration of what must be or what ought to be – which is referred to as 'morals' or 'ethics'.

(*Note*: 'Morals' and 'ethics': what difference is there

between these terms? The simplest answer is: none whatsoever. The term 'morals' derives from the Latin word for 'manners, customs', and 'ethics' derives from the Greek term for 'manners, customs'. They are therefore perfectly synonymous. Having said this, some philosophers have assigned different meanings to the two terms. In Kant, for example, 'morals' designates the ensemble of first principles, and 'ethics' refers to their application. Other philosophers refer to 'morals' as the theory of duties towards others, and to 'ethics' as the doctrine of salvation and wisdom. Indeed, there is no reason why different meanings should not be assigned to these terms, but, unless I indicate otherwise, I shall use them synonymously in the following pages.)

If philosophy, like religion, has its deepest roots in human 'finiteness' – the fact that for us mortals time is limited, and that we are the only beings in this world to be fully aware of this fact – it goes without saying that the question of what to do with our time cannot be avoided. As distinct from trees, oysters and rabbits, we think constantly about our relationship to time: about how we are going to spend the next hour or this evening, or the coming year. And sooner or later we are confronted – sometimes due to a sudden event that breaks our daily routine – with the question of what we are doing, what we should be doing, and what we must be doing with our lives – our time – as a whole.

This combination of the fact of mortality with our awareness of mortality contains all the questions of philosophy. The philosopher is principally not someone who believes that we are here as 'tourists', to amuse ourselves. Even if he does come to believe that amusement alone is worth experiencing, it will at least be the result of a process of thought, a reflection rather than a reflex. This thought process has three distinct stages: a *theoretical* stage, a *moral* or *ethical* stage, and a crowning conclusion as to *salvation* or *wisdom*.

The first task of philosophy is that of *theory*, an attempt to gain a sense of the world in which we live. Is it hostile or friendly, dangerous or docile, ordered or chaotic, mysterious or intelligible, beautiful or ugly? Any philosophy therefore takes as its starting point the natural sciences which reveal the structure of the universe – physics, mathematics, biology, and so on – and the disciplines which enlighten us about the history of the planet as well as our own origins. 'Let no one ignorant of geometry enter here,' said Plato to his students, referring to his school, the Academy; and thereafter no philosophy has ever seriously proposed to ignore scientific knowledge. But philosophy goes further and examines the *means* by which we acquire such knowledge. Philosophy attempts to define the nature of knowledge and to understand its methods (for example, how do we establish the causes of a natural phenomenon?) and its limits (for example, can one prove, yes or no, the existence of God?).

These two questions – the nature of the world, and the instruments for understanding it at our disposal as humans – constitute the essentials of the *theoretical* aspect of philosophy.

Besides our knowledge of the world and of its history, we must also interest ourselves in other people – those with whom we are going to share this existence. For not only are we not alone, but we could not be born and survive without the help of others, starting with our parents. How do we co-exist with others, what rules of the game must we learn, and how should we conduct ourselves – to be helpful, dignified and 'fair' in our dealings with others? This question is addressed by the second part of philosophy; the part which is not theoretical but practical, and which broadly concerns *ethics*.

But why should we learn about the world and its history, why bother trying to live in harmony with others? What is the point of all this effort? And does it have to make sense? These questions, and some others of a similar nature, bring us to the third dimension of philosophy, which touches upon the ultimate question of *salvation* or *wisdom*. If philosophy is the 'love' (*philo*) of 'wisdom' (*sophia*), it is at this point that it must make way for wisdom, which surpasses all philosophical understanding. To be a sage, by definition, is neither to aspire to wisdom or seek the condition of being a sage, but simply to live wisely, contentedly and as freely as possible, having finally overcome the fears sparked in us by our own finiteness.

I am aware this is becoming rather abstract, so I would like to offer some examples of the three aspects I have touched upon – theory, ethics and the quest for salvation or wisdom – in action.

The best course is therefore to plunge into the heart of the matter, to begin at the beginning; namely the philosophical schools which flourished in Greek antiquity. Let's consider the case of the first of the great philosophical movements, which passes through Plato and Aristotle to find its most perfected – or at least its most 'popular' – form in Stoicism. This is our way into our subject, after which we can explore the other major epochs in philosophy. We must also try to understand why and how men pass from one model of reality to another. Is it because the accepted version no longer satisfies, no longer convinces? After all, several versions of reality are inherently plausible.

You must understand that philosophy is an art not of questions but rather of answers. And as you are going to judge these things *for yourself* – this being another crucial promise of philosophy, because it is not religion, because it is not answerable to the truth of an Other – you will quickly see how profound these answers have been, how gripping, and how inspired.

2
'THE GREEK MIRACLE'

Most historians agree that philosophy first saw the light of day in Greece, some time around the sixth century BC. So sudden and so astonishing was its manifestation, it has become known as 'the Greek miracle'. But what was available, philosophically speaking, before the sixth century and in other civilisations? Why this sudden breakthrough?

I believe that two straightforward answers can be offered. The first is that, as far as we know, in all civilisations prior to and other than Greek antiquity, religion was a substitute for philosophy. An almost infinite variety of cults bears witness to this monopoly of meaning. It was in the protection of the gods, not in the free play of reason, that men traditionally sought their salvation. It also seems likely that the partially democratic nature of the political organisation of the city-state played some role in 'rational' investigation becoming emancipated from religious belief. Among the Greek elite, unprecedented freedom and autonomy of thought were favoured, and in their assemblies, the citizens acquired the habit of uninterrupted public debate, deliberation and argument.

Thus, in Athens, as early as the fourth century BC, a number of competing philosophical schools came to exist. Usually they were referred to by the name of the place where they first established themselves: Zeno of

Citium (*c.* 334–262 BC), the founding father of the Stoic school, held forth beneath colonnades covered with frescoes (the word 'stoicism' derives from the Greek word *stoa* meaning 'porch').

The lessons dispensed by Zeno beneath his famous 'painted porch' were open and free to all-comers. They were so popular that, after his death, the teachings were continued and extended by his disciples. His first successor was Cleanthes of Assos (*c.* 331–230 BC) followed by Chrysippus of Soli (*c.* 280–208 BC). Zeno, Cleanthes and Chrysippus are the three great names of what is called 'Early Greek Stoicism'. Aside from a short poem, the *Hymn to Zeus* by Cleanthes, almost nothing survives of the numerous works written by the first Stoics. Our knowledge of their philosophy comes by indirect means, through later writers (notably Cicero). Stoicism experienced a second flourishing, in Greece, in the second century BC, and a third, much later, in Rome. The major works of this third Roman phase no longer come down by word of mouth from Athenian philosophers succeeding each other at the head of the school; rather they come from a member of the imperial Roman court, Seneca (*c.* 8 BC–AD 65), who was also a tutor and advisor to Nero; from Musonius Rufus (AD 25–80) who taught Stoicism at Rome and was persecuted by the same Nero; from Epictetus (*c.* AD 50–130), a freed slave whose oral teachings were faithfully transmitted to posterity by his disciples – notably by Arrian, author of two works which were to travel down the ages, the *Discourses* and the *Enchiridion* or *Manual of Epictetus* (the title was said to derive from the fact that the maxims of Epictetus should be at every moment 'to hand' for those wanting to learn how to live – 'manual' coming from the Latin *manualis*, 'of or belonging to the hand'); and lastly, this body of Stoic teaching was disseminated by the Emperor Marcus Aurelius himself (AD 121–180).

I would now like to show you how a particular philosophy – in this case Stoicism – can address the challenge of human salvation quite differently to religions; how it can try to explain the need for us to conquer the fears born of our mortality, by employing the tools of reason alone. I shall pursue the three main lines of enquiry – theory, ethics and wisdom – outlined earlier. I shall also make plenty of room for quotations from the writers in question; while quotations can slow one down a little, they are essential to enable you to exercise your critical spirit. You need to get used to verifying for yourself whether what you are told is true or not, and for that, you need to read the original texts as early on as possible.

Theory, or the Contemplation of a Cosmic Order

To find one's place in the world, to learn how to live and act, we must first obtain knowledge of the world in which we find ourselves. This is the first task of a philosophical 'theory'.

In Greek, this activity calls itself *theoria*, and the origins of the word deserve our attention: *to theion* or *ta theia orao* means 'I see (*orao*) the divine (*theion*)' or 'divine things' (*theia*). And for the Stoics, *the-oria* is indeed a striving to contemplate that which is 'divine' in the reality surrounding us. In other words, the primary task of philosophy is to perceive what is *intrinsic* about the world: what is most real, most important and most meaningful. Now, in the tradition of Stoicism, the innermost essence of the world is *harmony*, *order* – both true and beautiful – which the Greeks referred to by the term *kosmos*.

If we want to form a simple idea of what was meant by *kosmos*, we must imagine the whole of the universe as if it were both ordered and animate. For the Stoics, the structure of the world – the cosmic order – is not merely magnificent, it is also comparable to a living being. The material world, the entire universe, fundamentally resembles a gigantic animal, of which each element – each organ – is conceived and adapted to the harmonious functioning of the whole. Each part, each member of this immense body, is perfectly in place and functions impeccably (although disasters *do* occur, they do not last for long, and order is soon restored) in the most literal sense: without fault, and in harmony with the other parts. And it is this that *theoria* helps us to unravel and understand.

In English, the term *cosmos* has resulted in, among other words, 'cosmetic'. Originally, this science of the body beautiful related to justness of proportions, then to the art of make-up, which sets off that which is 'well-made' and, if necessary, conceals that which is less so. It is this order, or *cosmos*, this ordained structure of the universe in its entirety that the Greeks named 'divine' (*theion*), and not – as with the Jews and Christians – a Being apart from or external to the universe, existing prior to and responsible for the act of its creation.

It is this *divinity*, therefore (nothing to do with a

personal Godhead), inextricably caught up with the natural order of things, that the Stoics invite us to contemplate (*theorein*), for example, by the study of sciences such as physics, astronomy or biology, which show the universe *in its entirety* to be 'well-made': from the regular movement of the planets down to the tiniest organisms. We can therefore say that the structure of the universe is not merely 'divine' and perfect of itself, but also 'rational', consonant with what the Greeks termed the *Logos* (from which we derive 'logic' and 'logical'), which exactly describes this admirable order of things. Which is why our human reason is capable of understanding and fathoming reality, through the exercise of *theoria*, as a biologist comes to comprehend the function of the organs of a living creature he dissects.

For the Stoics, opening one's eyes to the world was akin to the biologist examining the body of a mouse or a rabbit to find that everything therein is perfectly 'well-made': the eye admirably adapted for 'seeing well', the heart and the arteries for pumping blood through the entire body to keep life going; the stomach for digesting food, the lungs for oxygenating the muscles, and so on. All of which, in the eyes of the Stoic, is both 'logical' and 'divine'. Why divine? Not because a personal God is responsible for these marvels, but because these marvels are ready-made. Nor are we humans in any sense the inventors of this reality. On the contrary, we merely discover it.

It is here that Cicero, one of our principal sources for understanding the thought of the early Stoics, intervenes, in his *On the Nature of the Gods*. He scorns those thinkers, notably Epicurus, who think the world is not a *cosmos*, an order, but on the contrary a chaos. To which Cicero retorts:

> Let Epicurus mock as much as he likes . . . It remains no less true that nothing is more perfect than this world, which is an animate being, endowed with awareness, intelligence and reason.

This little excerpt gives us a sense of just how remote this way of thinking is from our own. If anyone claimed today that the world is alive, animate – that it possesses a soul and is endowed with reason – he would be considered crazy. But if we understand the Ancients correctly, what they are trying to say is by no means absurd: they were convinced that a 'logical' order was at work behind the apparent chaos of things and that human reason was able to discern the divine character of the universe.

It was this same idea, that the world possesses a soul of sorts, like a living being, which would later be termed 'animism' (Latin *anima*, meaning *soul*). This 'cosmology' (or conception of the *cosmos*) was also described as 'hylozoism', literally meaning that matter (*hyle*) is analogous to what is animal (*zoon*): that it is alive, in other words. The same doctrine would also be described by the term 'pantheism' (the doctrine that nature and the physical universe are constituents of the essence of God; from Greek *pan*, 'all', and *theos*, meaning 'God'): that all is God, since it is the totality of the universe that is divine, rather than there being a God beyond the world, creating it by remote control, so to speak. If I dwell on this vocabulary it is not out of a fondness for philosophical jargon (which often impresses more than it enlightens), but rather to enable you to approach these great philosophical texts for yourself, without grinding to a halt whenever you encounter these supposedly 'technical' terms.

From the point of view of Stoic *theoria*, then – and ignoring those temporary manifestations known as catastrophes – the *cosmos* is essentially *harmonious*. And, as we shall see, this would have important consequences for the 'practical' sphere (moral, legal and political). For if nature as a whole is harmonious, then it can serve as a model for *human* conduct, and the order of things *must* be just and good, as Marcus Aurelius insists in his *Meditations*:

> 'All that comes to pass comes to pass with justice.' You will find this to be so if you watch carefully. I do not mean only in accordance with the ordered nature of events, but in accordance with justice and as it were by someone who assigns to each thing its value. (IV.10)

What Marcus Aurelius suggests amounts to the idea that nature – when it functions normally and aside from the occasional accidents and catastrophes that occur – renders justice finally to each of us. It supplies to each of us our essential needs as individuals: a body which enables us to move about the world, an intelligence which permits us to adapt to the world, and natural resources which enable us to survive in the world. So that, in this great cosmic sharing out of goods, each receives his due.

This theory of justice ushers in what served as a first principle of all Roman law: 'to render to each what is his due' and to assign each to his proper place (which assumes, of course, that for each person and thing there is such a thing) – what the Greeks thought of as a 'natural place' in the *cosmos*, and that this *cosmos* was itself just and good.

You can see how, in this perspective, one of the ultimate aims of a human life is to find its rightful place within the cosmic order. For the majority of Greek thinkers – with the exception of the Epicureans whom we shall discuss later – it was through the pursuit of this quest, or, better, its accomplishment, that we attain happiness and the good life. From a similar perspective, the *theoria* itself implicitly possesses an aesthetic dimension, since the harmony of the universe which it reveals to us becomes for humans a model of beauty. Of course, just as there are natural catastrophes which seem to invalidate the idea of a good and just *cosmos* – although we are told that these are never more than temporary aberrations – so too there exist within nature things that are at first sight ugly, or even hideous. In their case, we must learn how to go beyond first impressions, the Stoics maintain, rather than remain content with appearances. Marcus Aurelius makes the point forcefully in his *Meditations*:

> The lion's wrinkled brow, the foam flowing from the boar's mouth, and many other phenomena that are far from beautiful if we look at them in isolation, do nevertheless because they follow from Nature's processes lend those a further ornament and fascination. And so, if a man has a feeling for, and a deeper insight into the processes of the Universe, there is hardly any of these but will somehow appear to present itself pleasantly to him . . . Even an old man or old woman will be seen

> to possess a certain perfection, a bloom, in the eyes of the sage, who will look upon the charms of his own boy slaves with sober eyes. (III, 2)

This is the same idea already expressed by one of the greatest Greek philosophers and model for the Stoics, Aristotle, when he denounced those who judge the world to be evil, ugly or disjointed: because they are looking only at a detail, without an adequate intelligence of the whole. If ordinary people think, in effect, that the world is imperfect, it is because, according to Aristotle, they commit the error 'of extending to the universe as a whole observations which bear only upon physical phenomena, and then only upon a small proportion of these. In fact, the physical world that surrounds us is the only one dominated by generation and corruption, but this world does not, one might say, constitute even a small part of the whole: so that it would be fairer to absolve the physical world in favour of the celestial world, than to condemn the latter on account of the former.' Naturally, if we restrict ourselves to examining our little corner of the cosmos, we shall not perceive the beauty of the whole, whereas the philosopher who contemplates, for example, the admirably regular movement of the planets will be able to raise himself to a higher plane through an understanding of the perfection of the whole, of which we are but an infinitesimal fragment.

Thus, the divine nature of the world is both *immanent and transcendent*. Again, I have used these philosophical terms because they will be useful to us later. Something that is *immanent* can be found nowhere else other than in this world. We say it is *transcendent* when the contrary applies. In this sense, the Christian God is transcendent in relation to the world, whereas the divine according to the Stoics, which is not to be located in some 'beyond' – being none other than the harmonious structure, cosmic or cosmetic, of the world as it is – is wholly immanent. Which does not prevent Stoic divinity from being defined equally as 'transcendent': not in relation to the world, of course, but in relation to man, given that it is *radically superior and exterior to him*. Men may discover it – with amazement – but in no sense do they invent it or produce it.

Chrysippus, the student of Zeno who succeeded Cleanthes as the third head of the Stoic school notes: 'Celestial things and those whose order is unchanging cannot be made by men.' These words are reported by Cicero, who adds in his commentary on the thought of the early Stoics:

> Wherefore the universe must be wise, and nature which holds all things in its embrace must excel in the perfection of reason [*Logos*]; and therefore the universe must be a God, and all the force of the universe must be held together by nature, which is divine. (*On the Nature of the Gods*, II, 11, 29–30)

We can therefore say of the divine, according to the Stoics, that it represents 'transcendence within immanence'; we can grasp the sense in which *theoria* is the contemplation of 'divine things' which, for all that they do not exist elsewhere than in the dimension of the real, are nonetheless entirely foreign to human activity.

I would like you to note again a difficult idea, to

which we shall return in more detail: the *theoria* of the Stoics reveals that which is most perfect and most 'real' – most 'divine', in the Greek sense – in the universe. In effect, what is most real, most essential, in their account of the cosmos, is its ordonnance, its harmony – and not, for example, the fact that at certain moments it has its defects, such as monsters or natural disasters. In this respect, *theoria*, which shows us all of this and gives us the means to understand it, is at once an 'ontology' (a doctrine which defines the innermost structure or 'essence' of being), and also a theory of knowledge (the study of the intellectual means by which we arrive at this understanding of the world).

What is worth trying to understand, here, is that philosophical *theoria* cannot be reduced to a specific science such as biology, astronomy, physics or chemistry. For, although it has constant recourse to these sciences, it is neither experimental, nor limited to a particular branch or object of study. For example, it is not interested solely in what is alive (like biology), or in the heavenly bodies (like astronomy), nor is it solely interested in inanimate matter (like physics); on the other hand it tries to seize the essence or inner structure of the world as a totality. This is ambitious, no doubt, but philosophy is not a science among other sciences, and even if it does take account of scientific findings, its fundamental intent is *not of a scientific order*. What it searches for is a meaning in this world and a means of relating our existence to what surrounds us, rather than a solely objective (scientific) understanding.

However, let us leave this aspect of things to one side for the time being. We shall return to it later when we need to define more closely the difference between philosophy and the exact sciences. I hope that you will sense already that this *theoria* – so different to our modern sciences and their supposedly 'neutral' principles, in that they describe what is and not what ought to be – must have practical implications in terms of morality, legality and politics. How could this description of the *cosmos* not have had implications for men who were asking themselves questions as to the best way of leading their lives?

Ethics: a System of Justice Based on Cosmic Order

What kind of ethics corresponds to the *theoria* that we have sketched so far? The answer is clear: one which encourages us to adjust and orientate ourselves to the *cosmos*, which for the Stoics was the watchword of all just actions, the very basis of all morals and all politics. For justice was, above all, *adjustment* – as a cabinetmaker shapes a piece of wood within a larger structure, such as a table – so our best efforts should be spent in striving to adjust ourselves to the harmonious and just natural order of things revealed to us by *theoria*. Knowledge is not entirely disinterested, as you see, because it opens directly onto ethics. Which is why the philosophical schools of antiquity, contrary to what happens today in schools and universities, placed less emphasis on speech than on actions, less on concepts than on the *exercise of wisdom*.

I will relate a brief anecdote so that you might fully understand the implications. Before Zeno founded the Stoic school, there was another school in Athens, from which the Stoics drew a great deal of their inspiration: that of the Cynics. Today the word 'cynic' implies something negative. To say that someone is 'cynical' is to say that he believes in nothing, acts without principles, doesn't care about values, has no respect for others, and so on. In antiquity, in the third century before Christ, it was a very different business, and the Cynics were, in fact, the most exacting of moralists.

The word has an interesting origin, deriving directly from the Greek word for 'dog'. What connection can there be between dogs and a school of philosophical wisdom? Here is the connection: the Cynics had a fundamental code of behaviour and strived to live according to nature, rather than according to artificial social conventions which they never stopped mocking. One of their favourite activities was needling the good citizens of Athens, in the streets and market squares, deriding their attitudes and beliefs – playing shock-the-bourgeois, as we might say today. Because of this behaviour they were frequently compared to those nasty little dogs who nip your ankles or start barking around your feet as if to deliberately annoy you.

It is also said that the Cynics – one of the most eminent of whom, Crates of Thebes, was Zeno's teacher – forced their students to perform practical exercises, encouraging them to discount the opinions of others in order to focus on the essential business of living in harmony with the cosmic order. They were told, for example, to drag a dead fish attached to a piece of string across the town square. You can imagine how the unhappy man forced to carry out this prank immediately found himself the target of mockery and abuse.

But it taught him a lesson or two! First, not to care for the opinions of others, or be deflected from pursuing what Cynic believers described as 'conversion': not conversion to a god, but to the cosmic reality from which human folly should never deflect us.

And, another more outrageous example: Crates occasionally made love in public with his wife Hipparchia. At the time, such behaviour was profoundly shocking, as it would be today. But he was acting in accordance with what might be termed 'cosmic ethics': the idea that morality and the art of living should borrow their principles from the harmonic law which regulates the entire *cosmos*. This rather extreme example suggests how *theoria* was for the Stoics a discipline to acquire, given that its practical consequences could be quite risky!

Cicero explains this cast of mind lucidly when summarising Stoic thought in another of his works, *On Moral Ends*:

> The starting-point for anyone who is to live in accordance with nature is the universe as a whole and its governance. Moreover, one cannot make correct judgements about good and evil unless one understands the whole system of nature, and even of the life of the gods, not know whether or not human nature is in harmony with that of the universe. Similarly, those ancient precepts of the wise that bid us to 'respect the right moment', to 'follow God', to 'know thyself', and 'do nothing to excess' cannot be grasped in their full force (which is immense) without a knowledge of physics. This science alone can reveal to us the power of nature to foster justice, and preserve friendship and other bonds of affection. (III, 73)

In which respect, according to Cicero, nature is 'the best of all governments'. You may consider how very different this antique vision of morality and politics is to what we believe today in our democracies, in which it is the will of men and not the natural order that must prevail. Thus we have adopted the principle of the majority to elect our representatives or make our laws. Conversely, we often doubt whether nature is even intrinsically 'good': when she is not confirming our worst suspicions with a hurricane or a tsunami, nature has become for us a neutral substance, morally indifferent, neither good nor bad.

For the Ancients, not only was nature before all else good, but in no sense was a majority of humans called upon to decide between good and evil, between just and unjust, because the criteria which enabled those distinctions all stemmed from the natural order, which was both external to and superior to men. Broadly speaking, the good was what was in accord with the cosmic order, *whether one willed it or not*, and what was bad was what ran contrary to this order, *whether one liked it or not*. The essential thing was to act, situation-by-situation, moment-by-moment, in accordance with the harmonious order of things, so as to find our proper place, which each of us was assigned within the Universal.

If you want to compare this conception of morality to something familiar and current in our society, think of ecology. For ecologists – and in this sense their ideas are akin to aspects of ancient Greek thought, without their necessarily realising it – nature forms a harmonious totality which it is in our interest to respect and even to imitate. In this sense the ecologists' conception of the 'biosphere', or of 'ecosystems', is close in spirit to that of the *cosmos*. In the words of the German philosopher Hans Jonas, a great theorist of contemporary ecology, 'the ends of man are at home in nature'. In other words, the objectives to which we ought to subscribe on the ethical plane are already inscribed, as the Stoics believed, in the natural order itself, so that our duty – the moral imperative – is not cut off from being, from nature as such.

As Chrysippus said, more than two thousand years before Hans Jonas, 'there is no other or more appropriate means of arriving at a definition of good or evil things, virtue or happiness, than to take our bearings from common nature and the governance of the universe', a proposition which Cicero in turn related in these terms: 'As for man, he was born to contemplate [*theorein*] and imitate the divine world . . . The world has virtue, and is also wise, and is consequently a Deity.' (*On the Nature of the Gods* II, 14).

Is this, then, the last word of philosophy? Does it reach its limits, in the realm of theory, by offering 'a vision of the world', from which moral principles are then deduced and in agreement with which humans should act? Not in the slightest! For we are still only on the threshold of the quest for salvation, of that attempt to raise ourselves to the level of true wisdom by abolishing all fears originating in human mortality, in time's passage, in death itself. It is only now, therefore, on the basis of a theory and a *praxis* (the translation of an idea into action; the practical side of an art or science, as distinct from its theoretical side) that we have just outlined, that Stoic philosophy approaches its true destination.

From Love of Wisdom to the Practice of Wisdom

Why bother with a *theoria*, or even an ethics? What is the point, after all, in taking all this trouble to contemplate the order of the universe, to grasp the innermost essence of being? Why try so doggedly to adjust ourselves to the world? No one is obliged to be a philosopher ... And yet it is here that we touch on the deepest question of all, the ultimate end of all philosophy: the question of salvation.

As with all philosophies, there is for the Stoics a realm 'beyond' morality. To use philosophers' jargon, this is what is termed 'soteriology', from the Greek *soterios* which means, quite simply, 'salvation'. As I have already suggested, this presents itself in relation to the fact of death, which leads us, sooner or later, to wonder about the irreversible nature of time and, consequently, about the best use we can make of it. Even if all humans do not become philosophers, all of us are one day or another affected by philosophical questions. As I have suggested, philosophy, unlike the great religions, promises to help us to 'save' ourselves, to conquer our fears, not through an Other, a God, but through our own strength and the use of our reason.

As the philosopher Hannah Arendt noted in *Between Past and Future* (1961), the Ancients, even before the birth of philosophy, traditionally found two ways of taking up the challenge of the inescapable fact of human mortality; two strategies, if you like, of attempting to outflank death, or at least, of outflanking the fear of death.

The first, quite naturally, resides in the simple fact of procreation: by having children, humans assure their 'continuity': becoming in a sense a part of the eternal cycle of nature, of a universe of things that can never die. The proof lies in the fact that our children resemble us physically as well as mentally. They carry forwards, through time, something of us. The drawback, of course, is that this way of accessing eternity really only benefits the species: if the latter appears to be potentially immortal as a result, the individual on the other hand is born, matures and dies. So, by aiming at self-perpetuation through the means of reproduction, not only does the individual human fall short, he fails to rise above the condition of the rest of brute creation. To put it plainly: however many children I have, it will not prevent me from dying, nor, worse still, from seeing them die before me. Admittedly, I will do my bit to ensure the survival of the species, but in no sense will I save the individual, the person. There is therefore no true salvation by means of procreation.

The second strategy was rather more elaborate: it consisted of performing heroic and glorious deeds to become the subject of an epic narrative, the *written trace* having as its principal virtue the conquest of transitory time. One might say that works of history – and in ancient Greece there already flourished some of the greatest historians, such as Thucydides and Herodotus – by recording the exceptional deeds accomplished by certain men, saved them from the oblivion which threatens everything that does not belong to the realm of nature.

Natural phenomena are cyclical. They repeat themselves indefinitely: night follows day; winter follows autumn; a clear day follows a storm. And this repetition guarantees that they cannot be forgotten: the natural world, in a peculiar but comprehensible way, effortlessly achieves a kind of 'immortality', whereas 'all things that owe their existence to men, such as works, deeds and words, are perishable, infected as it were, by the mortality of their authors' (Arendt). It is precisely this empire of the perishable, which glorious deeds, at least in theory, allowed the hero to combat. Thus, according to Hannah Arendt, the ultimate purpose of works of history in antiquity was to report 'heroic' deeds, such as the behaviour of Achilles during the Trojan war, in an attempt to rescue them from the world of oblivion and align them to events within the natural order:

> If mortals succeeded in endowing their works, deeds and words with some permanence and in arresting their perishability, then these things would, to a degree at least, enter and be at home in the world of everlastingness, and mortals themselves would find their place in the cosmos, where everything is immortal except men. ('The Concept of History, Ancient and Modern', in *Between Past and Future*, 1961)

This is true. In certain respects – thanks to writing, which is more stable and permanent than speech – the Greek heroes are not wholly dead, since we continue today to read accounts of their exploits. Glory can thus seem to be a form of personal immortality, which is no doubt why it was, and continues to be, coveted by so many. Although one must add that, for many others, it will never be more than a minor consolation, if not a form of vanity.

With the coming of philosophy, a third way of confronting the challenge of human mortality declared itself. I have already remarked how fear of death was, according to Epictetus – and all the great cosmologists – the ultimate motive for seeking philosophical wisdom. According to the Stoics, the sage is one who, thanks to a just exercise of thought and action, is able to attain a human version – if not of immortality – then at least of eternity. Admittedly, he is going to die, but death will not be for him the absolute end of everything. Rather it will be a transformation, a 'rite of passage', if you like, from one state to another, within a universal order whose perfection possesses complete stability, and by the same token possesses divinity.

We are going to die: this is a fact. The ripened corn will be harvested; this is a fact. Must we then, asks Epictetus, conceal the truth and refrain superstitiously from airing such thoughts because they are 'ill omens'? No, because 'ears of wheat may vanish, but the world remains'. The way in which this thought is expressed is worth our contemplation:

> You might just as well say that the fall of leaves is ill-omened, or for a fresh fig to change into a dried one, and a bunch of grapes into raisins. For all these changes are from a preceding state into a new and different state; and thus not destruction, but an ordered management and governance of things. Travelling abroad is

> likewise, a small change; and so is death, a greater change, from what presently is – and here I should not say: a change into what is not, but rather: into what presently is not. – In which case, then, shall I cease to be? – Yes, you will cease to be what you are, but become something else of which the universe then has need. (Epictetus, *Discourses,* III, 24, 91–4)

Or, according to Marcus Aurelius: 'You came into this world as a part: you will vanish into the whole which gave you birth, or rather you will be gathered up into its generative principle by the process of change.' (*Meditations*, IV,14)

What do such texts mean? They mean simply this: that having reached a certain level of wisdom, theoretical and practical, the human individual understands that death does not really exist, that it is but a passage from one state to the next; not an annihilation but a different state of being. As members of a divine and stable *cosmos*, we too can participate in this stability and this divinity. As soon as we understand this, we will become aware simultaneously how unjustified is our fear of death, not merely subjectively but also – in a pantheistic sense – objectively. Because the universe is eternal, we will remain for ever a fragment – we too will never cease to exist!

To arrive at a proper sense of this transformation is, for Epictetus, the object of all philosophical activity. It will allow each of us to attain a good and happy existence, by teaching us (according to the beautiful Stoic formula), 'to live and die like a God' – that is, to live and die as one who, perceiving his privileged connection with all other beings inside the cosmic harmony, attains a serene consciousness of the fact that, mortal in one sense, he is no less immortal in another. This is why, as in the case of Cicero, the Stoic tradition tended to 'deify' certain illustrious men such as Hercules or Aesculapius: these men, because their souls 'survived and enjoyed immortality, were rightly regarded as gods, for they were of the noblest nature and also immortal'.

These were the words of Cicero in *On the Nature of the Gods*. We might almost say that, according to this ancient concept of salvation, there are degrees of death: as if one died more or less, depending on whether one displayed more or less wisdom or 'illumination'. From this perspective, the good life was one which, despite the disappointed acknowledgement of one's finiteness, maintained the most direct possible link with eternity; in other words, with the divine ordinance to which the sage accedes through *theoria* or *contemplation*.

But let us first listen to Plato, in this lengthy passage from the *Timaeus*, which evokes the sublime power of man's sovereign faculty, his intellect (*nous*):

> God gave this sovereign faculty to be the divinity in each of us, being that part which, as we say, dwells at the top of the body, and inasmuch as we are a plant not of an earthly but of a heavenly growth, raises us from the earth to our kindred who are in heaven. For the divinity suspended the head and root of us from that place where the generation of the soul first began, and thus made the whole body upright. Now when a man gives himself over to the cravings of desire and

> ambition, and is eagerly striving to satisfy them, all his thoughts necessarily become mortal, and, as far as it is possible altogether to become such, he must become entirely mortal, because he has cherished his mortal part. But he who has been earnest in the love of knowledge and of true wisdom, and has exercised his intellect more than any other part of him, must have thoughts immortal and divine, if he attain truth, and in so far as human nature is capable of sharing in immortality, he must altogether be immortal. (90b–c)

And must also achieve a higher condition of happiness, adds Plato. To attain a successful life – one which is at once good and happy – we must remain faithful to the divine part of our nature, namely our intellect. For it is through the intellect that we attach ourselves, as by 'heavenly roots', to the divine and superior order of celestial harmony: 'Therefore must we attempt to flee this world as quickly as possible for the next; and such flight is to become like God, to the extent that we can. And becoming like God is becoming just and wholesome, by means of intellect.' (*Theaetetus*, 176a–b).

And we find a comparable statement in one of the most noted passages of Aristotle's *Nicomachean Ethics*, where he too defines the good life, 'the contemplative life', the only life which can lead us to perfect happiness, as a life by which we escape, at least in part, the condition of mere mortality. Some will perhaps claim that

> such a life is too rarefied for man's condition; for it is not in so far as he is man that he can live so, but in so far as something divine is present in him . . . If

> reason is divine, then, in comparison with man, the life according to reason is divine in comparison with human life. So we must not follow those who advise us, being human, to think only of human things, and, being mortal, of mortal things; but must, so far as we can, make ourselves immortal, and strain every nerve to live in accordance with what is best in us.' (X, 7)

Of course, this objective is by no means easy, and if philosophy is to be more than mere aspiration to wisdom – a genuine conquering of our fears – then it must be embodied in practical exercises.

Even though I am not myself a Stoic by inclination and am not convinced by this way of philosophical thinking, I must acknowledge the grandeur of its project and the formidable set of answers which it tries to bring. I would like to look at these now, by evoking a few of the exercises in wisdom to which Stoicism opens the way. For philosophy, as the word itself indicates, is not quite wisdom but only the love (*philo*) of wisdom (*sophia*). And, according to the Stoics, it is through practical exercise that one passes from one to the other. These exercises are intended to eradicate the anxiety associated with mortality – and in this respect they still retain, in my view, an inestimable value.

A Few Exercises in Wisdom

These almost exclusively concern our relation to time, for it is in the folds of time that these anxieties establish themselves, generating remorse and nostalgia for the past, and false hopes for the future. The exercises are all the more interesting and significant in that we encounter them time and again throughout the history of philosophy, in the thought of philosophers who are in other respects quite distant from the Stoics – in Epicurus and Lucretius, but also, curiously, in Spinoza and Nietzsche, and even in traditions remote from Western philosophy, such as Tibetan Buddhism. I will restrict myself to four examples.

The Burden of the Past and the Mirages of the Future

Let us begin with the essentials: in the eyes of the Stoics, the two great ills which prevent us from achieving fulfilment are nostalgia and hope, specifically attachment to the past and anxiety about the future. These block our access to the present moment, and prevent us from living life to the full. It has been said that Stoicism here anticipated one of the most profound insights of psychoanalysis: that he who remains the prisoner of his past will always be incapable of 'acting and enjoying', as Freud said; that the nostalgia for lost paradises, for the joys and sorrows of childhood, lays upon our lives a weight as heavy as it is unknown to us.

Marcus Aurelius expresses this conviction, perhaps better than anyone else, at the beginning of Book XII of his *Meditations*:

> It is in your power to secure at once all the objects which you dream of reaching by a roundabout route,

> if you will be fair to yourself: if you will leave all the past behind, commit the future to Providence, and direct the present alone, towards piety and justice. To piety, so that you may be content with what has been assigned to you – for Nature designed it for you and you for it; to justice, that you may freely and without circumlocution speak the truth and do those things that are in accord with law and in accord with the worth of each. (XII.1)

To be saved, to attain the wisdom that surpasses all philosophy, we must school ourselves to live without vain fears or pointless nostalgias. Once and for all we must stop living in the dimensions of time past and time future, which do not exist in reality, and adhere as much as possible to the present:

> Do not let your picture of the whole of your life confuse you, do not dwell upon all the manifold troubles which have come to pass and will come to pass; but ask yourself in regard to every passing moment: what is there here that cannot be borne and cannot be endured? Then remind yourself that it is not the future or the past that weighs heavy upon you, but always the present, and that this gradually grows less. (*Meditations,* VIII, 36)

Marcus Aurelius is quite insistent on this point: 'Remember that each of us lives only in the present moment, in the instant. All the rest is the past, or an uncertain future. The extent of life is therefore brief.' This is what we must confront. Or as Seneca expresses it, in the *Letters to Lucilius*: 'You must dispense with these two things: fear of the future, and the recollection of ancient ills. The latter no longer concerns me, the former has yet to concern me.' To which one might add, for good measure, that it is not only 'ancient ills' that spoil the present life of the unwise, but perversely and perhaps to a greater degree, the recollection of happy days irrevocably lost and which will return 'never more'.

If should now be clear why, paradoxically (and contrary to popular opinion), Stoicism would teach its disciples to part ways with those ideologies that promote the virtue of hope.

'Hope a Little Less, Love a Little More'

As one contemporary philosopher, André Comte-Sponville, has emphasised, Stoicism here is very close to one of the most subtle tenets of Oriental wisdom, and of Tibetan Buddhism in particular: contrary to the commonplace idea that one 'cannot live without hope', hope is the greatest of misfortunes. For it is by nature an absence, a lack, a source of tension in our lives. For we live in terms of plans, chasing after objectives located in a more or less distant future, and believing that our happiness depends upon their accomplishment.

What we forget is that there is no other reality than the one in which we are living here and now, and that this strange headlong flight from the present can only end in failure. The objective accomplished, we almost invariably experience a puzzling sense of indifference, if not disappointment. Like children who become bored with their toys the day after Christmas, the possession of things so ardently coveted makes us neither better nor happier than before. The difficulties of life and the tragedy of the human condition are not modified by ownership or success and, in the famous phrase of Seneca, 'while we wait for life, life passes'.

Perhaps you like imagining what you would do if you were to win the lottery: you would buy this and that; you would give some of it to this friend or that cousin; you would definitely give some of it to charity; and then you would take off on a trip around the world. And then what? In the end, it is always the gravestone that is silhouetted against the horizon, and you come to realise soon enough that the accumulation of all imaginable worldly goods solves nothing (although let us not be hypocrites: as the saying goes, money certainly does make poverty bearable).

Which is also why, according to a celebrated Buddhist proverb, you must learn to live as if this present moment were the most vital of your whole life, and as if those people in whose company you find yourself were the most important in your life. For nothing else exists, in truth: the past is no longer and the future is not yet. These temporal dimensions are real only to the imagination, which we 'shoulder' – like the 'beasts of burden' mocked by Nietzsche – merely to justify our incapacity to embrace what Nietzsche called (in entirely Stoic mode) *amor fati*: the love of reality for itself. Happiness lost, bliss deferred, and, by the same token, the present receding, consigned to nothingness whereas it is the only true dimension of existence.

It is with this perspective that the *Discourses* of Epictetus aimed to develop one of the more celebrated themes of Stoicism: namely, that the good life is a life stripped of both hopes and fears. In other words, a life reconciled to what is the case, a life which accepts the world as it is. As you can see, this reconciliation cannot sit alongside the conviction that the world is divine, harmonious and inherently good.

Here is how Epictetus puts the matter to his pupil: you must chase from your 'complaining' spirit

> all grief, fear, desire, envy, malice, avarice, effeminacy and intemperance. But these can be expelled only by looking to God, and attaching yourself to him alone, and consecrating yourself to his commands. If you wish for anything else, you will only be following what is stronger than you, with sighs and groans, always seeking happiness outside yourself, and never able to find it: for you seek it where it is not, and neglect to seek it where it is. (*Discourses*, II, 16, 45–7)

This passage must of course be read in a 'cosmic' or pantheistic sense, rather than in a monotheistic sense (*monotheism*: the belief in only one God).

Let us be very clear about this: the God of whom Epictetus speaks is not the personal God of Christianity, but merely an embodying of the *cosmos*, another name for the principle of universal reason which the Greeks named the *Logos*: the true face of destiny, that we have no choice but to accept, and should yearn for with our entire soul. Whereas, in fact, victims as we are of commonplace illusions, we keep thinking that we must oppose it so as to bend it to our purposes. As the master advises his pupil, once more:

> We must bring our own will into harmony with whatever comes to pass, so that none of the things which happen may occur against our will, nor those which do not happen be wished for by us. Those who have settled this as the philosopher's task have it in their power never to be disappointed in their desires, or fall prey to what they wish to avoid, but to lead personal lives free from sorrow, fear and perturbation. (*Discourses*, II, 14, 7–8)

Of course, such advice seems absurd to ordinary mortals: amounting to an especially insipid version of fatalism. This sort of wisdom might pass for folly, because it is based upon a vision of the world which requires a conceptual effort out of the ordinary to be grasped. But this is precisely what distinguishes philosophy from ordinary discussion, and, to me, why it possesses an irreplaceable charm.

I am far from being an advocate of Stoic resignation, and later on, when we touch upon contemporary materialism, I will explain more fully why this is so. However, I admire the fact that – when things are going well! – Stoicism can seem to offer a form of wisdom. There are moments when we seem to be here not to transform the world, but simply to be part of it, to experience the beauty and joy that it offers to us. For example, you are in the sea, scuba diving, and you put on your mask to look at the fish. You are not there to change things, to improve them, or to correct them; you are there to admire and accept things. It is somewhat in this spirit that Stoicism encourages us to reconcile ourselves to what is, to the present as it occurs, without hopes and regrets. Stoicism invites us to enjoy these moments of grace, and, to make them as numerous as possible, it suggests that we change ourselves rather than the order of things.

To move on from this concept to another essential Stoic counsel: because the only dimension of reality is the present, and because, of its nature, the present is in constant flux, it is wise for us to cultivate indifference or non-attachment to what is transient. Otherwise we store up the worst sufferings for ourselves.

Non-attachment

Stoicism, in a spirit remarkably close to that of Buddhism, appeals for an attitude of 'non-attachment' towards the things of this world. The Tibetan masters would no doubt have approved of this text from Epictetus:

> The principal and highest form of training, and one that stands at the very entrance to happiness, is, that when you become attached to something, let it not be as to something which cannot be taken away, but rather, as to something like an earthenware pot or crystal goblet, so that if it breaks, you may remember what kind of thing it was and not be distressed. So in this, too, when you kiss your child, or your brother, or your friend, never give way entirely to your affections, nor free rein to your imagination; but curb it, restrain it, like those who stand behind generals when they ride in triumph and remind them that they are

> but men. Remind yourself likewise that what you love is mortal, that what you love is not your own. It is granted to you for the present, and not irrevocably, not for ever, but like a fig or a bunch of grapes in the appointed season . . . What harm is there while you are kissing your child to murmur softly, 'Tomorrow you will die'? (*Discourses*, III, 24 84–8)

Let us be clear about what Epictetus is saying: it is not in any sense a case of being indifferent, as we might know it, and even less of lacking in the obligations which compassion imposes upon us in respect of others and, most importantly, of those close to us. He is saying that we must distrust all attachments that make us forget what the Buddhists call 'impermanence': the fact that nothing is stable in this world, that everything passes and changes, and that not to understand this is to create for oneself a hopelessness about what is past and a hope of what is yet to come. We must learn to content ourselves with the present, to love the present to the point of desiring nothing else and of regretting nothing whatsoever. Reason, which is our guide and which invites us to live in accordance with the harmony of the cosmos, must therefore be purified of that which weighs it down and falsifies it, whenever it strays into the unreal dimensions of time past and time future.

But once the truth of this is grasped we are still far from putting it into practice. Which is why Marcus Aurelius invites his disciples to embody it practically:

> So, if you separate, as I say, from this governing self [i.e. the mind] what is attached to it by passions, and what

> of time is left to run or has already flown, and make yourself like the sphere of Empedocles, 'rounded, rejoicing in the solitude which is about it', and practise only to live the life you are living, that is the present, then it will be in your power at least to live out the time that is left until you die, untroubled and dispensing kindness, and reconciled with your own good daemon. (*Meditations,* XII, 3)

As we shall see, this is precisely what Nietzsche refers to in his suggestive phrase, 'the innocence of becoming'. To attain this level of wisdom, we must have the courage to live our lives under the guidance of the 'future perfect' tense.

'When Catastrophe Strikes, I Will Be Ready'

What might this mean? Epictetus is speaking about his child, and what is at stake is once again death and the victories that philosophy can enable us to gain over (fear of) death. It is in this sense that the most practical of exercises connect to the most exalted spirituality. To live in the present and detach oneself from the regrets and anguish that define the past and the future is indeed to savour each moment of existence as it merits; in the full awareness that, for us mortals, it may be our last.

> Your time is circumscribed, and unless you use it to attain calm of mind, time will be gone and you will be gone and the opportunity to use it will not be yours again . . . Perform each action in life as though it were your last. (Epictetus, *Meditations,* II, 4, 5)

What is at issue spiritually in this exercise, where the subject shakes off all attachments to past and future, is therefore clear. It is a question of conquering the fears associated with our mortality, thanks to the use of an intuition that is not intellectual but intimate and almost physical.

There are moments of grace in our lives, instants when we have the rare experience of being completely reconciled to the world. Just now I gave the example of swimming underwater. Perhaps this doesn't mean anything to you or seems an odd choice, but I am sure you can imagine for yourself many other examples: a walk in a forest, a sunset, being in love, the calm and yet heightened state of something accomplished well – any of these experiences. In each case, we experience a feeling of serenity, of being at one with the world in which we find ourselves, where harmony occurs of its own accord, without being forced, so that time seems to stop, making room for the enduring present, a present which cannot be undermined by anything in the past or future.

To see to it that life as a whole resembles such moments: that is the fundamental project of Stoic wisdom. It is at this point that we touch on something resembling salvation, in the sense that nothing further can trouble a serenity which comes from the extinguishing of fears concerning other dimensions of time. When he achieves this degree of enlightenment, the sage does indeed live 'like a god', in the eternity of an instant that nothing can diminish.

From which you can understand how, for Stoicism as for Buddhism, the tense in which the struggle against anxiety is to be waged is indeed the 'future perfect'. In effect: 'When destiny strikes, I shall have been prepared for it.' When catastrophe – be it illness, poverty or death, all the ills linked to the irreversible nature of time – *will have taken place*, I shall be able to confront it thanks to the ability I have acquired to live in the present. In other words one can love the world as it is, no matter what transpires:

> If some so-called 'undesirable' event should befall you, it will in the first place be an immediate relief to you that it was not unexpected . . . You will say to yourself, 'I knew all along that I am mortal. I knew that in this life I might have to go away, that I might be cast into exile. I knew that I might be thrown into prison.' Then if you reflect within yourself and ask from what quarter the accident has come, you will at once remember that it comes from the region of things outside our will, which are not ours. (Epictetus, *Discourses,* III, 105–6)

This wisdom still speaks to us today, through the centuries and overarching many cultures. However, we no longer inhabit the world of Greek antiquity, and the great cosmologies have for the most part vanished, together with the 'wisdom of the ages'. This raises an important question: why and how do we pass from one vision of the world to another? Or, in other words, why are there different philosophies which seem to follow on from one another in the history of ideas, rather than a single system of thought which survives the passage of time and suffices us once and for all?

Let's examine this question in detail through looking at the most recent example: that of the doctrines of salvation associated with the great cosmologies. Why was Stoic wisdom not enough to stifle the emergence of competing systems of thought, and, specifically, to prevent the spread of Christianity? After all, Christianity was to deal Stoicism a lethal blow, relegating it to a marginal position for centuries.

By taking a specific example of how one vision of the world yields to another, we may learn lessons of a more general kind about the development of philosophy. As far as Stoicism goes, we recognise that, however grandiose the positions it advocated, a major weakness affected its response to the question of salvation – one which was to leave room for a competing version to establish itself, and which consequently allowed the machine of history to set off again.

As you have probably noticed, the Stoic doctrine of salvation is resolutely *anonymous* and *impersonal*. It promises us eternity, certainly, but of a non-personal kind, as an oblivious fragment of the *cosmos*: death, for the Stoic, is a mere rite of passage, which involves a transition from a state of individual consciousness – you and I, as living and thinking beings – to a state of oneness with the *cosmos*, in the course of which we lose everything that constitutes our self-awareness and individuality. It is by no means certain, therefore, that this doctrine can fully answer the questions raised by our anxiety about human finiteness. Stoicism tries valiantly to relieve us of the fears linked to death, but at the cost of obliterating our individual identity. What we would like above all is to be reunited with our loved ones, and, if possible, with their voices, their faces – not in the form of undifferentiated cosmic fragments, such as pebbles or vegetables. In this arena, Christianity might be said to have used its big guns. It promises us no less than everything that we would wish for: personal immortality *and* the salvation of our loved ones. Exploiting what it saw as a weakness in Greek wisdom, Christianity created a new doctrine of salvation so 'effective' it opened a chasm in the philosophies of Antiquity and dominated the Occidental world for nearly fifteen hundred years.

3
THE VICTORY OF CHRISTIANITY OVER GREEK PHILOSOPHY

When I was a student – in 1968, when religious questions were not the most fashionable – we basically ignored the medieval frame of mind. In other words, we lumped together and cheerfully channel-hopped our way through the great monotheist religions. It was possible to pass our exams and even become a philosophy professor by knowing next to nothing about Judaism, Islam or Christianity. Of course, we had to attend lectures on ancient thought – Greek thought, above all – after which we could cut straight to Descartes. Without any transition, we leapt fifteen centuries, broadly speaking from the end of the second century (the late Stoics) to the beginning of the seventeenth century. As a result, for years I knew more or less nothing about the intellectual history of Christianity, beyond the cultural commonplaces.

This strikes me as absurd, and I would not wish you to repeat this mistake. Even if one is not a believer, and all the more so if one is hostile to religion – as we shall see in the case of Nietzsche – we have no right to ignorance. If only to oppose it, we must at least be familiar with religion in its various forms, and understand what we are opposing. At the least, it explains many facets of the world in which we live, which is the direct product of a religious world-view. There is not a museum of art, even of contemporary art, which does not require a minimum of theological understanding, if one is to fully understand its contents; and there is no single conflict in the world today that is not more or less linked to the history of religious communities: Catholics and Protestants in Northern Ireland; Muslims, Orthodox and Catholics in the Balkans; Animists, Christians and Islamists in Africa, and so on.

Yet, according to the definition of philosophy given at the start of this book, you would not normally expect it to include a chapter on Christianity. The notion of a 'Christian philosophy' might seem out of place and contradictory to what I have been proposing at length. Religion is the prime example of a non-philosophical quest for salvation – given its assumption of God and a need for faith – rather than by means of human reason. So, why discuss it here? For four simple reasons, which I will now set out briefly.

First, as I suggested at the end of the last chapter, the doctrine of Christian salvation, although fundamentally non-philosophical, even anti-philosophical, found itself in direct competition with Greek philosophy. It was to profit, so to speak, from the flaws which weakened the Stoic response to the question of salvation. The Christian solution even appropriated the vocabulary of philosophy for its own ends, assigning new religious meanings, and put forward an entirely fresh response to the question of our relation to death and to time. Its approach supplanted more or less entirely the answers supplied by the philosophy of the preceding centuries. This merits our attention.

The second reason is that even if the doctrine of Christian salvation is not a philosophy, there remains nonetheless a place for the exercise of reason at the heart of Christianity: on the one hand, to reflect on the great evangelical texts – to interpret the message of Christ; on the other hand, to gain an understanding of the natural order which, in so far as it is God's work, must surely bear some mark of its creator. We shall return to this question, but it will suffice for now to understand that, paradoxically, there was to be a place after all – subordinate and modest, certainly, but nonetheless real – for philosophical activity at the heart of Christianity: a role for human reason to clarify and reinforce a doctrine of salvation, even if the latter would remain fundamentally religious and founded on faith.

The third reason proceeds directly from the second: that there is no more illuminating way of understanding philosophy than to compare it with what it is not; to place it in relation to that to which it is most firmly opposed and yet most closely linked, namely religion. Ultimately setting their sights on the glittering prize of salvation, both religion and philosophy are closely linked, through their attempt to conquer anxiety over human mortality. They are at the same time opposed, because the means used by each are not merely different but irreconcilable. The Gospels, the Gospel of John in particular, reveal a level of familiarity with Greek philosophy, notably Stoicism. There can be no doubt, therefore, as to the confrontation and competition between opposing doctrines of salvation – Christian and Greek. An examination of the reason why the former prevailed over the latter is essential for an understanding not only of the nature of philosophy, but also for an understanding of how, after the long epoch during which Christian ideas were dominant, philosophy was able to re-emerge and set off for new horizons – those of modern thought.

Finally, there are in Christian thought, above all in the realm of ethics, ideas which are of great significance even today, and even for non-believers; ideas which, once detached from their purely religious origins, acquired an autonomy that came to be assimilated into modern philosophy. For example, the idea that the moral worth of a person does not lie in his inherited gifts or natural talents, but in the free use he makes of them, is a notion which Christianity gave to the world, and which many modern ethical systems would adopt for their purposes. It would be obtuse to try and pass from the Greek experience to modern philosophy without any mention of Christian thought.

I would like to explain why Christian thought gained the upper hand over Greek thought and dominated Europe until the Renaissance. This is no small achievement: there must surely be reasons for this hegemony. In fact, as we shall see, Christians came up with answers to human questions about mortality which have no equivalent in Greek thought – answers so 'successful', if you like, so 'attractive' and so indispensable that they convinced a large proportion of humanity.

To compare this doctrine of salvation and those philosophies of salvation which dispensed with God, I am going to follow once more the formula of theory, ethics and wisdom. To keep to essentials, I will first summarise the key characteristics which marked the radical rupture of Christianity with the Greek world – five characteristics which will allow you to understand how, based on a new *theoria*, Christianity was able to outline a new morality and a doctrine of salvation based on love. Thus did religion capture the hearts of men.

How Religion Replaced Reason with Faith

Firstly, and most fundamentally: the *Logos*, which as we as have seen for the Stoics merged with the impersonal, harmonious and divine structure of the *cosmos* as a whole, came to be identified for Christians with a single and unique personality, that of Christ. To the horror of the Greeks, the new believers maintained that the *Logos* – in other words the divine principle – was in no sense identical with the harmonious order of the world, but was incarnated in one outstanding individual, namely Christ.

Perhaps this distinction leaves you stone cold. After all, what does it matter – for us, today – that the *Logos* (for the Stoics a 'logical' ordering of the world) came to mean Christ as far as Christians were concerned? I might reply that today there exist more than a thousand million Christians – and that for this reason alone, to understand what drives them, their motives, the content and meaning of their faith, is not absurd for anyone with a modicum of interest in their fellow men. But this answer would be inadequate. For what is at stake in this seemingly abstract debate as to where the divine principle resides – whether in the structure of the universe or in the personality of one exceptional man – is no less than the transition from an anonymous and blind doctrine of salvation to one that promises not only that we shall be saved by one person, Christ, but that we shall be saved as individuals in our own right: for what we are, and as we are.

This 'personalising' of salvation allows us firstly to comprehend – by means of a concrete example – how mankind can pass from one vision of the world to another: how a new response to reality comes to prevail over an older response because it 'adds' something: a greater power of conviction, but also considerable advantages over what had preceded it. But there is more: by resting its case upon a definition of the human person and an unprecedented idea of love, Christianity was to have an incalculable effect upon the history of ideas. To give one example, it is quite clear that, in this Christian re-evaluation of the human person, of the individual as such, the philosophy of human rights to which we subscribe today would never have established itself. It is essential therefore that we have a more or less accurate idea of the chain of reasoning which led Christianity to break so radically with the Stoic past. And to have such an understanding, we must first grasp that in the vernacular translations of the Gospels which narrate the life of Jesus, the term *Logos* – borrowed directly from the Stoics – is translated by 'word'. For Greek thought in general, and for Stoicism in particular, the idea that the *Logos* could designate anything other than the rational (therefore true, therefore beautiful) order of the universe was unthinkable. In their eyes, to claim that a mere mortal could constitute the *Logos*, or 'the word incarnate', as the Gospels express it, was insanity. It was to assign the attribute of divinity to a mere human being, whereas the divine, as you will recall, is interchangeable with the universal cosmic order, and can in no sense be identified with a single puny individual, whatever his credentials.

The Romans – notably under Marcus Aurelius, Roman Emperor at the close of the second century and the last great Stoic thinker – did not hold back from massacring Christians on account of their intolerable 'deviance'. For this was a time when ideas were not playthings.

What exactly was at issue in this apparently innocent change in the meaning of a single word? The answer: nothing less than a revolution in the definition of divinity. And as we know, revolutions do not take place without suffering.

Let us return for a moment to the text in which John, author of the Fourth Gospel, effects this diversion of meaning away from the Stoic sense. Here is what he says – with my comments italicised inside brackets:

> In the beginning was the Word [*Logos*], and the Word was with God, and the Word was God . . . All things were made by him; and without him was not any thing made that was made. [*Up to this point, all is well, and the Stoics could still be in agreement with John, especially with the notion that the* Logos *and the divine are one and the same reality.*] And the Word was made flesh [*things start to take a turn for the worse!*] and dwelt among us [*quite unacceptable – the divine has become*

> *man, as incarnated in Jesus, none of which makes sense to a Stoic*]. And we beheld his glory, the glory as of the only begotten of the Father, full of grace and truth. [*sheer madness, for the Greek sages: the followers of Christ are now presented as witnesses of the transformation of the* Logos/Word *(or Godhead) – into Mankind (or Christ) as if the latter were son of the former.*] (1 John 1)

What is the meaning of this? To put it simply now – although at the time it was a matter of life or death – the divine had shifted ground: it was no longer an impersonal structure, but an extraordinary individual, in the form of Jesus, the 'Man-God'. This was an unfathomable shift, which was to direct European humanity along a quite different path than that set out by the Greeks. In a few lines of text, the very opening lines of his Gospel, John invites us to believe that the incarnate Word, the divine as such, no longer designates the rational and harmonious structure of the *cosmos*, the universal order as such, but refers instead to a simple individual.

We shall see how Marcus Aurelius would order the death of Saint Justin Martyr, a former Stoic who became the first Father of the Church and the first philosopher to convert to Christianity, but let us continue for a moment to explore the new aspects of this entirely original *theoria*. You will recall that *theoria* always comprises two aspects: on the one hand an unveiling of the essential structure of the universe (the divine); on the other hand the instruments of knowledge which it employs to arrive at this understanding (the vision or contemplation). Now it is not simply the divine, the *theion*, which is utterly changed here by becoming an individual being; but also the *orao*, the fashion of seeing, or act of contemplating, understanding and approaching reality that is transformed. From now on, it is no longer *reason* that will be the theoretical faculty par excellence, but *faith*. In which respect, religion will soon declare its opposition to the rationality at the heart of philosophy, and, by these means, depose philosophy itself.

And so, faith begins to supplant reason. For Christians, truth is no longer accessed through the exercise of a human reason which can grasp the rational and 'logical' order of the cosmic totality by virtue of its being an eminent component of that same order. From now on, what will permit man to approach the divine, to know it and to contemplate it, belongs to a quite different order. What will count here, above all, is no longer intelligence but *trust* in the word of a man, the Man-God, Christ, who claims to be the son of God, the *Logos* incarnate. We are going to believe Him, because He is worthy of this act of faith – and the miracles He accomplishes will play their part in the credit which is accorded to Him.

You will recall that trust originally meant 'faith'. To contemplate God, the appropriate *theoretical* instrument is faith, not reason, and this means placing all our confidence in the words of Christ announcing the 'good news': according to which we shall be saved by faith and not by 'works'; in other words, our all too human actions, however admirable these might be. It is no longer a case of *thinking for oneself*, but rather of *placing trust in another*. And in that, no doubt, lies the most profound and significant difference between philosophy and religion.

From which proceeds the importance of *bearing witness*, as the First Epistle of Saint John makes clear:

> That which was from the beginning, which we have heard, which we have seen with our eyes, which we have looked upon, and our hands have handled, of the Word [*Logos*] of life – for the life was manifested, and we have seen it, and bear witness, and shew unto you that eternal life, which was with the Father, and was manifested unto us – that which we have seen and heard declare we unto you, that you also may have fellowship with us. (1 John 1)

Of course, it is of Christ that John is speaking, and you will see that his words rest upon a quite different logic to that of reflection and reason: it is not a case of *arguing* for or against the existence of a God – such a topic for argument goes beyond human reason – but a case of bearing witness and believing, of declaring that we have seen 'the Word made flesh', Christ; that we have 'handled', touched, heard, spoken with Him, and that this witness is to be trusted. You are free to believe or not to believe that the divine *Logos*, the life eternal which was with the Father, has been incarnated in a Man-God who came down to Earth. But it is no longer a case of working this out by intelligence and reason. If anything, the reverse is the case: 'Happy the poor in spirit', as Christ says in the Gospels, for they will believe and consequently see God. Whereas the 'confident', the 'haughty' – as Augustine described the philosophers – will walk past the truth in all the finery of their pride and arrogance.

Third: what is required to put into practice the new

theoria is not the comprehension of philosophers, but the humility of simple folk. It is no longer a question of thinking for oneself but of believing in and through another. The theme of humility is omnipresent in the critiques of the two greatest Christian philosophers: St Augustine, who lived in the Roman Empire in the fourth century after Christ, and Pascal, who lived in seventeenth-century France. Each based their attack on philosophy (which they never missed an opportunity to criticize, to the point that it seems for them to have been the great enemy) on the fact that it was an exercise of pride.

There is no shortage of passages from St Augustine denouncing the pride and vanity of philosophers who refused to accept that Christ could be the incarnation of the Word, of the divine principle and who could not tolerate the modesty of a Godhead reduced to the status of a humble mortal, vulnerable to suffering and death. As he says in *The City of God*, taking aim at philosophers: 'The haughty disdained to accept this God as their master, because "the Word was made flesh and dwelt amongst us".' This was intolerable to philosophers. Why? Because it required that they hang up their intelligence and their reason in the church vestibule to make room for faith and belief.

There is, then, a double humility in religion, which opposes it to Greek philosophy from the outset, and which corresponds to the two aspects of the *theoria*, that of the divinit (*theion*) and that of contemplative seeing (*orao*). On the one hand there is the humility, 'objective' if you like, of a *divine Logos* which finds itself 'reduced' in the person of Jesus to the status of a lowly mortal (too lowly, for the Greeks). On the other hand, there is the subjective humility of our being enjoined by believers to 'let go' of our own thinking faculty, to forsake reason for trust, so as to make place for faith. Nothing is more significant in this respect than the terms employed by Augustine in *The City of God*:

> Swollen with pride by the high opinion they had of their science, they [philosophers] did not hear Christ when he said: 'Learn of Me, because I am meek, and humble of heart, and you shall find peace.'

The founding text of Christianity, here, occurs in the New Testament, in the First Epistle to the Corinthians, written by St Paul. It is a difficult text, but it was to have such a profound influence on the subsequent history of Christianity, it demands to be read with some care. It shows how the idea of the incarnation of the Word – the idea, therefore, that the divine *Logos* was made man, and that Christ, in this sense, is the son of God – is unacceptable, as much for the Jews as for the Greeks: unacceptable to the Jews, because a diminished God, who lets himself be put to death on a cross without defending himself seems contemptible, and contrary to their image of an all-powerful and angry Jahweh; unacceptable to the Greeks, too, because an incarnation as mundane as this diminishes the grandeur of the *Logos* as conceived by the 'wisdom of the ages' of Stoic philosophy. Here is the text:

> Hath not God made foolish the wisdom of this world? For after that in the wisdom of God the world by wisdom knew not God, it pleased God by the foolishness of

> preaching to save them that believe. For the Jews require a sign, and the Greeks seek after wisdom. But we preach Christ crucified, unto the Jews a stumbling-block, and unto the Greeks foolishness; but unto them which are called, both Jews and Greeks, Christ the power of God, and the wisdom of God. Because the foolishness of God is wiser than men; and the weakness of God is stronger than men. (1 Corinthians 1: 20–25)

Here Paul traces the image, incredible at this time, of a God who is no longer bombastic: neither angry, nor terrifying, nor all-powerful, like the God of the Jews; rather he is meek and forgiving to the point of allowing himself to be crucified – which to the Jews of the time only went to show that he definitely had no divine attributes! Nor was this God cosmic and sublime, like the divinity of the Greeks, who identified God with the perfect structure of the entire universe. And yet it was through the humility of this new God, and His demanding humility of those who would follow Him, that he became the representative of the weak, the lowly, the excluded. Hundreds of millions of people recognised themselves, and still do so today, in the strange power of this very weakness.

According to believers, it was this, specifically, that the philosophers could not stomach. I would like to dwell on this for a moment, so that you can assess this theme of religious humility opposed to philosophical arrogance. The opposition is everywhere to be found in *The City of God* where Augustine takes a poke at the most important philosophers of his time (distant disciples of Plato, to be precise) who refuse to accept that the divine could become human. According to Augustine, their intelligence should have led them to the same conclusion as the Christians:

> But humility was the necessary condition for submission to this truth; and it is no easy task to persuade the proud necks of you philosophers to accept this yoke. For what is there incredible – especially for you who hold certain opinions which should encourage you to belief – what is there incredible in the assertion that God has assumed a human soul and body? . . . Why is it, then, that when the Christian faith is urged upon you, you straightaway forget, or pretend to have no knowledge of, your customary arguments and doctrines? What reason is there for your refusal to become Christians on account of opinions which are your own, though you yourselves attack them? It can only be that Christ came in humility, and that you are proud. (*The City of God*, X, 29)

This articulates the double-humility of which I spoke a moment ago: that of a God who agrees to 'abase himself' to the point of becoming a man amongst men; and that of the believer who renounces his reasoning to place all his trust in the word of Jesus, and thereby make room for faith.

As is now clear, the two aspects of Christian *theoria* – the definition of the divine and the definition of the intellectual attitude which allows contact with it – are poles apart from those of Greek philosophy. This leads us into the fourth characteristic.

Fourth: in a perspective which accords primacy to humility and to faith over reason – to 'thinking through an other' rather than 'thinking for oneself' – philosophy does not vanish entirely but becomes the 'handmaiden' to religion. This view appears first in the eleventh century, in the writings of Peter Damian, a Christian apologist close to the papacy. It had an immense impact because it indicated that, henceforth, in Christian doctrine, reason would be entirely subjected to the faith which guides it.

So, is there a Christian philosophy? The response must be 'yes' and 'no'. No, in the sense that the highest truths in Christianity, as in all of the major monotheistic religions, are termed 'revealed truths': that is, truths transmitted by the word of Christ, the son of God himself. These truths become an active belief system. We might then be tempted to say that there is no further role for philosophy within Christianity, because the essentials are decided by faith. However, one might also assert that in spite of everything there remains a Christian philosophical activity, although relegated to second place. Saint Paul emphasises repeatedly in his Epistles that there remains a dual role for reason and consequently for purely philosophical activity. On the one hand, Christ expresses himself in terms of symbols and parables (the latter in particular need interpreting, if we are to draw out their deeper sense). Even if the words of Christ have the distinction, a little like the great orally transmitted myths, legends and fairytales, of speaking to everyone, they do require the effort of reflection and intelligence to decipher their more hidden meanings.

But this is not simply a matter of interpreting the Scriptures. Nature too – 'the created order' – needs to be read; a rational approach to which must be capable of showing how it 'demonstrates' the existence of God through the beauty and goodness of His works. From St Thomas Aquinas onwards, in the thirteenth century, this aspect of Christian philosophy was to become more and more important. And it would lead to what theologians refer to as 'the proofs for the existence of God'; in particular, the proof which shows that the world is perfectly constructed – the Greeks did not get *everything* wrong, after all!

You can see now why one might say that there both is and is not a Christian philosophy. There must clearly be a place for rational activity – to interpret Scripture and comprehend the natural order sufficiently to draw the correct conclusions as to the Christian divinity. But the doctrine of salvation is no longer the prerogative of philosophy, and, even if they do not in principle contradict one another, the truths revealed by faith take precedence over those deduced by reason.

This leads us to the fifth and last characteristic: no longer the master of the doctrine of salvation, philosophy must become 'scholastic'; a dry discipline and not a body of wisdom or a living principle. This point is crucial, for it explains why, even today, at a time when many people think they have definitely left behind the Christian era, the majority of philosophers continue to reject the idea that philosophy can be a doctrine of salvation, or even an apprenticeship to wisdom. At school as at university, philosophy has become essentially the history of ideas, a purely 'discursive' apprenticeship, contrary to what it had been in ancient Greece.

With Christianity this rupture was introduced, whereby the Greek philosopher ceased to invite his disciple to practise those exercises in wisdom which were the basis of teaching in the academies. This is quite understandable, since the doctrine of salvation, founded on faith and on revelation, no longer belonged to the domain of reason. Philosophy for the most part evolved into a learned commentary upon realities which transcended philosophy and were removed from its sphere of practice: one philosophises about the meaning of the Scriptures, or about nature as a work of God, but not about the ultimate ends and purposes of human life. Even today, it seems that philosophy starts from and speaks about realities exterior to itself: the philosophy of science, of law, of language, of politics, of art, of morals and so on, but almost never *philo-sophia*: the love of wisdom. With a few rare exceptions, contemporary philosophy still assumes the secondary status to which it was relegated by the victory of Christianity over Greek thought. Personally, I find this regrettable – I shall try to explain why in the chapter devoted to contemporary philosophy.

But for the present, let us trace how Christianity would also evolve a new ethics which was in several respects at odds with the Greeks' consensus.

The Birth of the Modern Idea of Humanity

One might have expected that the stranglehold of religion over thought would have as a consequence a reduction of the ethical plane. However, one could argue that the reverse happened. Christianity was to bring to ethical thought at least three novel ideas, none of which was Greek – or not *essentially* Greek – and all of which directly linked to the theoretical revolution we have just observed in action. These new ideas were arresting in their modernity. It is probably impossible for us, no matter how much effort we make, to imagine just how disruptive they must have seemed to contemporaries. The Greek world was fundamentally an aristocratic world, a universe organised as a hierarchy in which those most endowed by nature should in principle be 'at the top', while the less endowed saw themselves occupying inferior ranks. And we should not forget that the Greek city-state was founded on slavery.

In direct contradiction, Christianity was to introduce the notion that humanity was fundamentally identical, that men were equal in dignity – an unprecedented idea at the time, and one to which our world owes its entire democratic inheritance. But this notion of equality did not come from nowhere.

Here, I shall restrict myself to describing the three characteristics which are critical for an understanding of early Christian ethics. First: freedom of choice, 'free will', became the foundation of morals, and the notion of the equal dignity of all human beings made its first appearance. The natural (Greek) order is fundamentally hierarchical: for each category of beings, nature displays a full range, from the most sublime excellence to the deepest mediocrity. It is evident that if nature is our guide, we are endowed unequally: we are more or less strong, swift, tall, beautiful, intelligent etc. All natural gifts are unequally distributed.

In the moral vocabulary of the ancient Greeks, the notion of 'virtue' was always directly linked to those of talent or natural endowment. Which is why, to give a typical example of Greek thought, Aristotle can tranquilly speak of a 'virtuous eye' in one of his works devoted to ethics, by which he simply meant an 'excellent' eye, a perfectly functioning eye, neither long-sighted nor short-sighted.

To explain further: the Greek world is an aristocratic world, one which rests entirely upon the conviction that there exists a natural hierarchy, of organs of sight, of plants, or of animals, but also of men: some men are born to command, others to obey, which is why Greek political life accommodates itself easily to the notion of slavery.

For Christians, this belief in a natural hierarchy has no legitimacy. To speak of a 'virtuous' eye no longer makes any sense, because the gifts received at birth are unequally distributed among men; some men are much stronger or more intelligent than others, just as there exist in nature sharper eyes and less sharp eyes. These inequalities have no bearing on morals. Here all that counts is how we use the qualities with which we have been endowed, not the qualities themselves. What counts as moral or immoral is the act of choice, what philosophers began to call 'free will'. This may seem self-evident, but it was literally unheard-of at the time, and it turned an entire world-order upside down. To summarise: we exit an aristocratic universe and we enter a 'meritocratic' universe, a world which first and foremost values not natural or inherited qualities, but the *merit* which each of us displays in making use of them. We leave behind a natural order of inequality and enter a constructed order (in the sense that it is devised by us) of equality; human dignity is the same for everyone, whatever their actual inequalities, because it is connected to our freedom to choose how to act, not upon our innate endowments.

The Christian argument is at once very simple and very powerful. It says the following: there is indisputable proof that the talents bestowed by nature are not intrinsically virtuous, that they are in no sense inherently moral, because, without exception, they can be employed as much for ill as for good. Strength, beauty, intelligence – all natural gifts received at birth – are self-evidently qualities, but not on a moral plane. You can use your strength, your beauty or your intelligence to commit the most wicked crime, and you demonstrate by this alone that there is nothing inherently virtuous about natural gifts. Therefore, you can choose what use to make of them, whether good or bad, but it is the use that is moral or immoral, not the gifts themselves. 'Free will' becomes the determining factor of the morality of an action. With this idea, Christianity revolutionised the history of thought. For the first time in human history, liberty rather than nature had become the foundation of morality.

At the same time, the idea of the equal dignity of all human beings makes its first appearance: and Christianity was to become the precursor of modern democracy. Although at times hostile to the Church, the French Revolution – and, to some extent, the 1789 Declaration of the Rights of Man – owes to Christianity an essential part of its egalitarian message. We see today how civilisations that have not experienced Christianity have great difficulties in fostering democratic regimes, because the notion of equality is not so deep-rooted.

The second upheaval is directly linked to the first: that, in the moral sphere, the spirit is more important than the letter, the 'inner forum' of conscience more decisive than the 'outward forum' of secular law, which can never be more than an external imposition. Here, a passage from the Gospels may serve as a model: it concerns the famous episode where Christ comes to the defence of a woman accused of adultery, whom the crowd is preparing to stone to death. At this time adultery, the deception of a husband or a wife, was universally regarded as a sin, and the law stated that an adulteress should be stoned to death. But what about the spirit, the 'inner conscience'? Christ steps out from the God-fearing crowd and appeals directly to their conscience, saying

> In your heart of hearts (*inner forum*), are you sure that all is well? And were you to examine yourselves, are you certain that what you would find would be better than this woman whom you are preparing to kill and who, perhaps, has sinned only through love? He that is without sin among you, let him first cast a stone at her . . .

And all these men, instead of following the letter of the law, look into themselves, into their hearts, and reflect on their own defects. And they begin to doubt that they should act as merciless judges.

It is difficult at first to grasp the immense novelty of Christianity, not merely in relation to Greek thought, but even more so perhaps in relation to the Jewish world.

Because Christianity placed so much weight on conscience, on the spirit over the letter, it imposed almost no jurisdiction over everyday life. Rituals such as eating no fish on Fridays are mostly modern, dating back no further than the nineteenth century and having no origins whatsoever in the Gospels. You can read and re-read the Gospels, and find next to nothing about what you should or should not eat, how and to whom you should get married; there are hardly any rituals required for proving to yourself and others that you are a good and committed believer. While the lives of Orthodox Jews and Muslims are filled with duties to be carried out in civil society, Christianity left everything up to the individual as to whether something is good or not.

This attitude smoothed the passage to democracy, and the arrival of secular rather than religious societies: as morality was essentially a matter of internal conscience, it had less reason to come into conflict with external conventions. It mattered little whether one prayed once or a hundred times daily, or that one was forbidden to eat this or that: all laws, more or less, became acceptable if they did not infringe the spirit of the Christian message.

And now to the third fundamental innovation: the modern notion of humanity makes its entrance. Not, of course, that this notion was unknown to the Greeks, or to other civilisations: there existed an awareness of a 'human species', as distinct from other animal species – the Stoics in particular were especially attached to the idea that all men formed a single community. They were true 'cosmopolitans'.

But with Christianity, the idea of a common humanity

acquired a new strength. Based on the equal dignity of all human beings, it was to take on an ethical aspect. As soon as free will becomes the foundation of moral action and virtue is located not in natural, 'unequal' gifts, but in the use to which they are put, then it goes without saying that all men are of equal merit. Humanity would never again be able to divide itself (philosophically) according to a natural and aristocratic hierarchy of beings: between superior and inferior, gifted and less gifted, masters and slaves. From then on, according to Christians, we were all 'brothers', on the same level as creatures of God and endowed with the same capacity to choose whether to act well or badly. Rich or poor, intelligent or simple; it no longer holds any importance. And this idea of equality leads to a primarily ethical conception of humanity. The Greek concept of 'barbarian' – synonymous with 'stranger' ('anyone not Greek') – will slowly disappear to be replaced by the conviction that humanity is ONE. To conclude, we could say that Christianity is the first *universalist* ethos; *universalism* meaning the doctrine or belief in universal salvation.

In a wholly unprecedented manner, Christianity responded forcefully to the fundamental question of how to conquer the fears aroused in man by the sense of his own mortality. Whereas the Stoics represent death as a transition from a personal to an impersonal state of existence (from a condition of individual consciousness to that of a cosmic fragment without consciousness), the Christian version of salvation promises us nothing less than individual immortality. The idea of which is not easy to resist.

This promise is not superficial: on the contrary it is

part of a coherent intellectual framework – a concept of love and the resurrection of the body – and one of extraordinary profundity.

Salvation through Love

The heart of the Christian doctrine of salvation is directly linked to the transition from a cosmic to a personal conception of the *Logos*, of divinity as such. Its three most characteristic traits stem from this transformation, and it becomes clear how the Christian arguments came to prevail over the Stoic doctrine of salvation.

First: if the *Logos*, or divine principle, is incarnated in the person of Christ, the idea of providence changes its meaning. Instead of a blind and anonymous destiny, as with the Stoics, it becomes a personal and benevolent act, comparable to that of a father for his children. The salvation to which we can now aspire – based no longer on a cosmic order but on the commandments of this personal divinity – is personal. *Individual* immortality is promised to us. This turning point was described in 160 AD in a work by the first Father of the Church, Saint Justin. What is so unusual about this *Dialogue* is that it is written in a surprisingly familiar style, for its time: Justin was well versed in Greek thought and studied the Christian doctrine of salvation in relation to the great texts of Plato, Aristotle and the Stoics. He also describes how he has been variously a Stoic, an Aristotelian, a Pythagorean, and a fervent Platonist – before eventually becoming a Christian! His testimony is therefore extremely valuable to us and worth spending some time on.

Justin belonged to a group of early Christians known as the 'Apologists' and was their prime mover during the second century. At this time the persecution of Christians was still a feature of daily life in the Roman Empire. The first Christian theologians began to compile 'apologias' or 'reasoned defences' of their religion, which were addressed to the Roman emperors, in the hope of defending their community against hostile rumours about their form of worship. Christians were regularly accused of the most bizarre behaviour, for example, that they worshipped a God with the head of a donkey, indulged in cannibalistic sacrifice and ritual murders, or were involved in such debauched acts as incest. None of which was true.

The apologias compiled by Justin were intended to testify to the reality of Christian practices and to counter malicious gossip. The first apologia, dating from 150 AD, was sent to the Emperor Antoninus; the second to Marcus Aurelius, one of the greatest representatives of Stoic thought and also, curiously, a statesman. Roman law decreed that Christians could not be harrassed unless they were denounced by an individual 'of credibility'. It fell to a philosopher of the Cynic school named Crescens to take on this sinister role: a staunch adversary of Justin, and one who was jealous of his public repute. Crescens had Justin and six of his pupils condemned, and they were decapitated in AD 165 – under the reign of Marcus Aurelius. The transcript of Justin's trial has survived, the only primary document relating to the martyrdom of a Christian thinker in Rome during this period.

It is especially interesting to read what Justin professes, as he is confronted by Stoics intent on executing him. The bone of contention, unsurprisingly, concerned the doctrine of salvation. According to Justin, the Christian version of salvation wins out over that of the Stoics:

> They [the Greek thinkers] attempt moreover to persuade us that God takes care of the universe with its genera and species, but not of you and I, and each of us individually, since otherwise we would surely not need to pray to Him night and day! (*Dialogue with Trypho*, 1)

The implacable and blind Fate of the Ancients gives way to the benevolent wisdom of an individual who loves us as individuals, and in a way that no one else loves us. It is love that becomes the key to salvation. But, this is not love in the usual sense; it is what Christian thinkers will call 'love in God'.

This leads us to the second characteristic: love is stronger than death. What link can there be between the sentiment of love and the question of what can save us from mortality and death? It is simplest to start from the Christian proposition that there are, fundamentally, three faces of love, which between them form a coherent 'system'. First, there is the love that we might call 'love-as-attachment': in the sense that we are *bound* to another, to the point of not being able to imagine life without this other. We can experience this love as much within a family as with a lover. On this point, Christians were united with Stoics and Buddhists in viewing this love as the most dangerous and the least enlightened of all. Not only because it risks diverting us from our true duties towards God, but also because it cannot survive death and it cannot tolerate rupture and change. Aside from the fact that it is usually possessive and jealous, love-as-attachment stores up for us the worst of all sufferings – the loss of loved ones. At the opposite extreme is what we might call 'compassion': a love that drives us to care for strangers when they are in need. We still encounter this today, in the form of Christian charity, or, for example, the work of a humanitarian agency. And, finally, there is 'love-in-God'. Here and only here is the ultimate source of salvation, which, for Christians, will prove stronger than death.

Let us examine these definitions of love a little more closely. They are fascinating, because they have all endured for centuries and remain as active today as at the time when they first came into being.

You will remember that Stoicism regards the fear of death as the greatest obstacle to the happy life (likewise in Buddhism). And this anxiety is not without its connection to love. In simple terms there is an apparently insurmountable contradiction between love, which leads to attachment, and death, which leads to separation. If the law of this world is one of finiteness and mutability, and if, as the Buddhists maintain, everything is 'impermanent' – changing and perishable – then we sin by lack of wisdom if we attach ourselves to things or persons that are mortal. Not that we must resort to indifference, of course, which neither Stoic sage nor Buddhist monk would for a moment countenance: compassion and benevolence to others, indeed to all other forms of life, must remain the highest ethical imperative of our behaviour. But *passion* is not acceptable in the home of the wise man, and familial ties, when they become too binding, must be loosened. Which is why, like the Greek sage, the Buddhist monk lives, as much as possible, in a condition of solitude. (The word 'monk' derives from the Greek *monos*, meaning 'alone'.) It is truly in solitude that wisdom can bloom, uncompromised by the difficulties associated with all forms of attachment. It is impossible, in effect, to have a wife or husband, children or friends without becoming in some degree attached to them. We must free ourselves of these ties if we wish to overcome the fear of death. As Buddhist wisdom reminds us:

> The ideal condition in which to die is one where you have abandoned everything, inwardly and outwardly, so that there should be, at this crucial moment, the least possible longing, desire or attachment to which the soul can cling. This is why, before dying, we should free ourselves from all our goods, friends and family. (Sogyal Rinpoche, *The Tibetan Book of Living and Dying*)

Or, as the New Testament expresses it:

> For he that soweth to his flesh shall of the flesh reap corruption; but he that soweth to the Spirit shall of the Spirit reap life everlasting. (Epistle to the Galatians, VI, 8)

From the same perspective, Saint Augustine condemns those who attach themselves to mortal creatures through bonds of love:

> You seek a happy life in the region of death. How can there be a happy life where there is not even life? (*Confessions* IV, 12)

Similarly, Pascal, in his *Pensées* (1658–62), brilliantly elaborates the reasons why it is unworthy not only to attach oneself to others, but to allow another to attach himself or herself to one. I strongly recommend reading the whole of this profoundly important text:

> It is unjust that men should attach themselves to me, even though they do it with pleasure and voluntarily. I should deceive those in whom I evince this desire; for I am an end for no person, and have not the wherewithal to satisfy them. Am I not about to die? And thus the object of their attachment will die. Therefore, as I would be culpable in causing a falsehood to be believed, though I should employ gentle persuasion, though it should be believed with pleasure, and though it should give me pleasure; even so I am culpable in making myself loved. And if I attract persons to attach themselves to me, I should warn those who are ready to consent to such a lie that they should not believe it, whatever advantage I might derive from it; and likewise that they ought not to attach themselves to me; for they should be spending their life and their efforts in pleasing God, or in seeking Him. (*Pensées*, 471)

In the same vein, Augustine describes how, when he was a young man and still a pagan, he let his heart be broken by attaching himself to a friend who suddenly died. He believed that his grief was caused entirely by this lack of wisdom:

> The reason why this grief had penetrated me so easily and so deeply, was that I had poured my soul out onto quicksand by loving a person sure to die, as if he would never die. (*Confessions*, IV, 8)

He describes a human love as seeking in the other only those 'marks of affection' which increase our standing, reassure us and satisfy our own ego:

> Hence the mourning when a friend dies, the darkness of grief. And as the sweetness is turned to bitterness the heart is flooded with tears. The lost life of those who die becomes the death of those still living. (*Confessions*, IV, 9)

We must therefore learn how to resist exclusive attachments, since 'everything perishes in this world, everything is subject to failure and death'. As soon as it involves mortal creatures, we must ensure that

> our soul does not become stuck and glued to these transient things by loving them through the physical senses. For as these perishable creatures pass along the path of things that race towards non-existence, they rend the soul with pestilential desires, and torment it without cease; for the soul loves to be in them and take its repose among the objects of its love. But in these things there is no point of rest, for they are impermanent, they flee away and cannot be followed with the bodily senses. No one can fully grasp them even while they are present. (*Confessions*, IV, 10)

This is beautifully expressed, and it seems to me that the Stoic sage as well as the Buddhist would agree wholeheartedly with these words from a Christian convert.

On the other hand, who says man is mortal? The entire originality of the Christian message resides in 'the good news' of literal immortality – *resurrection*, in other words, and not merely of souls but of individual human bodies. If humans are immortal as long as they obey the commandments of God and if we suppose that this immortality is not merely compatible with earthly love but possibly one of its consequences, then why deprive ourselves? Why not become attached to our nearest and dearest, if Christ promises that we shall be reunited after our biological death?

Thus, between 'love-as-attachment' and love as simple universal compassion towards others, a place opens up for a third form of love: the love 'in' God of creatures who are themselves eternal. And it is here that Augustine wishes to lead us:

> Happy, my God, is the person who loves you, and his friend in you, and his enemy for your sake. Though left alone, he loses none who are dear to him; for all are dear in the one who cannot be lost. Who is that but God, our God . . . No one can lose you, my God, unless he abandons you. (*Confessions*, IV, 9)

To which we might add, that no one can lose the individuals he loves, unless he ceases to love them in God; in other words, ceases to love what is eternal in them, bound to God and protected by Him. This promise is, to say the least, tempting. And it was to find its most complete form in that ultimate statement of the Christian doctrine of salvation: that of resurrection, unique amongst all of the major religions.

To the third trait: personal immortality at last – the resurrection of the flesh as the culmination of the Christian doctrine of salvation. For the Buddhist, the individual is but an illusion, something destined for dissolution and impermanence; for the Stoic the individual self is destined to merge into the totality of the *cosmos*; Christianity on the contrary promises immortality of the individual person: his soul, his body, his face, his beloved voice – as long as he is saved by the grace of God. Now here is a seductive promise, since it is through love, and not only love of God, not only of one's neighbour, but most particularly love of one's nearest and dearest that salvation is to be gained. Thus does love become the solution for Christians.

This is why Augustine, having conducted a radical critique of 'love-as-attachment' in general, does not banish it when its object is divine – is God himself, and God's creatures:

> If souls please you, they are being loved in God; for they too are mutable and acquire stability by being established in him. Otherwise they go their way and perish . . . Stand with him and you will stand fast. (*Confessions*, IV, 12)

Nothing is mores striking than the serenity with which Augustine evokes the bereavements he has suffered, not prior to his conversion, but after his conversion – starting with the death of his mother, to whom he was very close:

> Then when she breathed her last, the boy Adeodatus cried out in sorrow and was pressed by all of us to be silent. In this way too something of the child in me, which had slipped towards weeping, was checked and silenced by the voice of reason. For we did not think it right to accompany her obsequies with tearful dirges

> and lamentations, since in most cases it is customary to use such mourning to imply sorrow for the miserable state of those who die, or even to assume their complete extinction. Whereas my mother's dying meant neither that her state was miserable nor that she was suffering extinction, of which we were confident because of the evidence of her virtuous life. (*Confessions*, IX, 12)

In the same way, Augustine does not hesitate to evoke 'the happy deaths of two friends', whom he also had the happiness of seeing converted and who consequently would benefit in turn from 'the resurrection of the just' (*Confessions*, IX, 3). As always, Augustine finds the apt word, for it is indeed the resurrection which ultimately founds this third kind of love – the love of God. Neither attachment to worldly things – which is doomed to endure the worst sufferings, on which Stoics and Buddhists agree – nor a vague compassion towards the much vaunted 'neighbour', meaning the world and his wife; but rather a love which is attached, physical and personal, towards other individuals, those nearest as well as neighbouring, provided that this love is founded 'in God', that is, in the context of a faith which makes real the possibility of resurrection.

From which emerges the direct link between love and the doctrine of salvation. For it is through love in God that Christ alone proves to be the one who, making 'death itself die' and 'making this mortal flesh put on immortality', promises that the life of our loves will not come to an end with earthly death.

We should not forget that the idea of personal immortality was already present in a number of philosophies and religions prior to Christianity, nonetheless, the Christian version of resurrection is unique in closely associating three fundamental themes for its doctrine of the happy life: that of the personal immortality of the soul, the resurrection of the body and of salvation through love. Without resurrection – significantly designated as 'the good news' in the Acts of the Apostles – the whole message of Christ collapses, as the New Testament makes unambiguously clear:

> Now if it be preached that Christ rose from the dead, how say some among you that there is no resurrection of the dead? But if there be no resurrection of the dead, then is Christ not risen: And if Christ be not risen, then is our preaching vain, and your faith is also vain. Yea, and we are found false witnesses of God; because we have testified of God that he raised up Christ: whom he raised not up, if so be that the dead rise not. (1 Corinthians 15: 12–15)

The resurrection is, so to speak, the alpha and omega of the Christian doctrine of salvation: it stands not only at the end of our earthly life, but equally so at the beginning, in the liturgy of the baptism, considered as a first death and symbolised as such by immersion in water, and as a first entrance to true life, one of a community wedded as individuals to eternity.

This cannot be emphasised too much: that it is not merely the soul that is resuscitated, but the 'soul-body' in its entirety; and therefore the individual. When Jesus reappears to his disciples after his death, he suggests – to remove all doubts – that they touch him, and, as proof of his 'materiality', he asks for a little food, which he eats before them:

> So that if the Spirit of Him that raised up Jesus from the dead dwell in you, He that raised up Christ from the dead shall also quicken your mortal bodies by his Spirit that dwelleth in you. (Romans, 8:11)

While it is difficult, even impossible, to imagine the resurrection of the flesh – With which body shall we be reborn, and at what age? What is meant by a 'spiritual body', a 'glorious' body, and so on?, and for all that this doctrine is one of the unfathomable mysteries of a Revelation which goes far beyond our powers of reason, even as Christians – the difficulty changes nothing. The teaching is entirely unambiguous.

Although atheists would have us believe otherwise, the Christian religion is not entirely given over to waging war against the body, the flesh, the senses. If that were so, how would Christianity have accepted that the divine principle be incarnated in the person of Christ, that the *Logos* take on the physical aspect of a simple mortal? Even the official catechism of the Church, perhaps not the most boldly original of texts, insists:

> The flesh is the hinge of salvation. We believe in God who is creator of the flesh; we believe in the Word made flesh in order to redeem the flesh; we believe in the resurrection of the flesh, the fulfillment of both the creation and the redemption of the flesh . . . We believe in the true resurrection of this flesh that we now possess. We sow a corruptible body in the tomb, but he raises

> up an incorruptible body, a spiritual body. (*Catechism of the Catholic Church*, 1015–17)

One can be a non-believer, but one cannot maintain that Christianity is a religion dedicated to contempt for the flesh. Because this is simply not the case.

Taking resurrection as the end-point of the doctrine of salvation, we can begin to understand what enabled Christianity to rule more or less unchallenged over philosophy for nearly fifteen hundred years.

The Christian response to mortality, for believers at least, is without question the most 'effective' of all responses: it would seem to be the only version of salvation that enables us not only to transcend the fear of death, but also to beat death itself. And by doing so in terms of individual identity, rather than anonymity or abstraction, it seems to be the only version that offers a truly definitive victory of personal immortality over our condition as mortals.

The personalising of the *Logos* changes all factors in the equation. If the promises made to me by Christ are genuine; and if divine providence takes me in hand as an individual, however humble, then my immortality will also, in turn, be personal. In which case, death itself is finally overcome, and not merely the fears it arouses in me. Immortality is no longer the anonymous and cosmic event proposed by Stoicism, but the individual and conscious resurrection of souls together with their 'glorious' bodies. In this sense, it is 'love in God' which confers its ultimate meaning upon this revolution effected by Christianity in relation to Greek thought. It is this new definition of love, found at the heart of the new doctrine of salvation, which finally turns out to be 'stronger than death'.

How and why did this doctrine begin to recede with the Renaissance? How and why did philosophy succeed in gaining the upper hand once more over religion, from the seventeenth century onwards? What was philosophy able to propose in its place? It is to the birth of modern philosophy we must now turn our attention.

4
HUMANISM, OR THE BIRTH OF MODERN PHILOSOPHY

Let us retrace our steps for a moment. We have seen how ancient philosophy founded a doctrine of salvation in terms of a consideration of the *cosmos*. In the eyes of a pupil of the Stoic schools, it went without saying that, to be saved – in other words, to overcome the fear of death – we must in the first place endeavour to understand the cosmic order; secondly, do our utmost to imitate it; and thirdly, merge ourselves in it, by finding our rightful place therein, and thus succeed in attaining a kind of eternity.

We have also analysed the ways in which Christian doctrine prevailed over Greek thought, and how, to attain salvation, a Christian was required to acknowledge the Word, through the humility of faith, observe the commandments, and finally to practise love in God in order to enter the kingdom of eternal life.

The modern world arose out of the collapse of ancient cosmology and a new questioning of religious authority, and eventually a scientific revolution unprecedented in the history of humanity, which occurred in Europe over the course of one hundred and fifty years. To my knowledge, no other civilisation has undergone such a radical upheaval in the fabric of its culture.

This upheaval began with the publication of Copernicus's work *On the Revolutions of the Heavenly*

Bodies in 1543, continued with that of Newton's *Principia Mathematica* in 1687, and took in Descartes' *Principles of Philosophy* (1644) and Galileo's *Dialogue Concerning the Two Chief World Systems* (1632). These four dates and these four authors were to mark the history of thought as no other thinkers before them. A new era was established, which, in many respects, we still inhabit today. It was not only man who 'lost his place' in the world, as is often said, but the *cosmos* itself – the enclosed and harmonious frame of human existence since antiquity – quite simply evaporated; leaving the intellects of the time in a state of confusion it is virtually impossible for us to imagine today.

Modern physics annihilated the foundations of the ancient world-picture – through its assertion, for example, that the world is not round, enclosed, hierarchical and divinely ordered, but rather is an infinite chaos devoid of sense; a field of forces and objects jostling for place without harmony – and weakened considerably the foundations of Christian religion.

Science called into question issues that the Church had unwisely adopted – the age of the Earth, its relation to the Sun, the date of birth of mankind and of animal species etc – and invited men to adopt an attitude of doubt and a critical spirit incompatible with respect for religious authority. Belief, at this time fettered in shackles rigidly imposed by the Church, started to waver, so that the most enlightened individuals found themselves dramatically at odds with ancient doctrines of salvation which were becoming less and less credible.

Nowadays we speak of a loss of bearings, together

with the suggestion that amongst the young in particular, things are falling apart – manners and knowledge, the sense of history, interest in politics, minimal acquaintance with literature, religion and art – but I would suggest that this harking back to 'the good old days' is as nothing compared to the disorientation men in the sixteenth and seventeenth centuries must have felt. This is why we speak of 'Humanism' in relation to this period: man found himself for the first time alone, deprived of the support of both *cosmos* and God.

To try to imagine the abyss which opened at this time, we need to put ourselves in the shoes of someone who is beginning to realise that the most recent scientific discoveries have just invalidated the idea of the *cosmos* as just and good; that in consequence it is going to be impossible for him to take the *cosmos* as his ethical model; and, for good measure, that the belief in God which might have served as his life-raft is taking in water! Our seventeenth-century friend is going to have to rethink, from scratch, the question of *theoria*, that of *ethical conduct* and that of *salvation*.

First, on the theoretical level: if the world is no longer finite, ordered and harmonious, and is instead infinite and chaotic, according to the new physics, how can he make sense of this world and his place in it? One of the greatest modern historians of science, Alexander Koyré, describes the scientific revolution of the sixteenth and seventeenth centuries so well, that I will simply quote his account here:

> the destruction of the idea of a *Cosmos*; that is, the disappearance, from philosophically and scientifically

> valid concepts, of the conception of the world as a finite, closed, and hierarchically ordered whole . . . and its replacement by an indefinite and even infinite universe which is bound together by the identity of its fundamental components and laws, and in which all these components are placed on the same level of being . . . The immediate effect of the Copernican revolution was to spread scepticism and bewilderment, to which the famous verses of John Donne (written in 1611) give such striking expression:
>
> . . . new Philosophy calls all in doubt,
> The Element of fire is quite put out;
> The Sun is lost, and th'Earth, and no man's wit
> Can well direct him where to looke for it.
> 'Tis all in pieces, all coherence gone;
> All just supply, and all Relation.
>
> (Alexander Koyré, *From the Closed World to the Infinite Universe, 11, 47)*

'All coherence gone': no harmonious *cosmos* and no natural moral order. How can we comprehend the anguish which must have possessed Renaissance men?

Second, in terms of ethics, this theoretical revolution has an effect as obvious as it is devastating: if the universe no longer has any of the attributes of a *cosmos*, it cannot serve as a model for imitation within the moral sphere. And if Christianity itself is unsure of its foundations, if obedience to God is no longer a given, where then are we to look for the principles of a new relationship between men, and a new foundation for the common life? We are going to have to rebuild the morality which has served as a model for centuries. Nothing less.

Third, the *doctrine of salvation*: you can see for yourself that, for the same reasons, neither the ancient model nor the Christian model remain credible for anyone of a critical and informed disposition.

The challenges taken up by modern philosophy on these three fronts were of an unprecedented scale and complexity – and urgency, too: as the verses by Donne suggest, never had humanity been so convulsed and at the same time rendered so resourceless, intellectually, morally and spiritually. But, as we shall see, the greatness of modern philosophy is to have been equal to these challenges.

A New Theory of Knowledge

As you may imagine, numerous factors played a role in the passage from a closed world to infinite space. Of key importance was technological progress, notably the development of new astronomical instruments such as the telescope, which enabled observations that could not be reconciled with the framework of existing and ancient models of cosmology. One example which made a strong impression on contemporaries was the discovery of the *novae* – new stars – or conversely, the disappearance of existing stars, neither of which conformed with the dogma of 'celestial immutability' so dear to the ancients. Their notion of the ultimate perfection of the *cosmos* resided in the fact that it was eternal and immutable, that nothing within it could change. For the Greeks, this orthodoxy represented something absolutely essential – human salvation depended on it – yet contemporary astronomers were revealing that this belief was false: quite simply, it was contradicted by the facts.

There were of course many other causes for the decline of the old cosmologies, notably economic and social, but the new scientific discoveries were the most critical. Before we can even begin to consider the upheavals which this eclipse of the cosmos was to cause within the moral sphere, we must understand that it was above all the *theoria* which had wholly changed its meaning and direction.

The book that was to underpin the whole of modern philosophy and one that remains a monument in the history of thought, was Emmanuel Kant's *Critique of Pure Reason* (1781). I am not going to attempt to summarise it here in a few sentences, but, although it is challenging work, I would like to try to give you an idea of how it came to reformulate in totally novel terms the question of the theoria.

If the world is no longer a *cosmos* but a chaos, a field of forces engaged in constant conflict with each other, it becomes clear that knowledge can no longer take the form of contemplation (*theoria*). One might say that, after the collapse of a beautiful cosmic order and its replacement by a nature devoid of sense and at war with itself, there is *nothing divine* about the universe. Order, harmony, beauty and goodness are no longer the first principles. Consequently, to re-establish a degree of coherence, so that the world in which men live continues to have a meaning, it was going to require man himself, or at least men of learning, to introduce some order into a universe which seemed no longer to offer any of its own. The new task of contemporary science was no longer to frame itself as the passive contemplation of a beauty inscribed beforehand in nature; it was to do a job of *work*, namely the active *construction* of laws which would endow a disenchanted universe with meaning. Science was no longer a passive spectacle; it was an activity of the mind.

I would like to give at least one example of this transition from passive to active knowledge, from assumption to construction, from ancient *theoria* to modern science. Let us consider the principle of *causality* – the principle according to which every effect has a cause or, if you prefer, every phenomenon must have a rational explanation. Instead of being content with discovering the order of the world through contemplation, the 'modern' philosopher or *savant* (scholar; learned person) would attempt to introduce, by means of a principle of causality, some coherence and sense into the chaos of natural phenomena. He would try *actively* to make logical connections between certain phenomena, which he was to consider as effects, and others which he succeeded in detecting as causes. In other words, thought was no longer a 'seeing', an *orao*, as the word 'theoria' suggests, but an 'acting', a work which consists in relating natural phenomena to each other, so that they form a chain of connections: and thus explain each other. This is what will come to be termed 'scientific method', virtually unknown as such to the ancients, and which would become the fundamental building block of modern science.

An example of this 'acting' is Claude Bernard, the great nineteenth-century French doctor and biologist, who published his celebrated *Introduction to the Study of Experimental Medicine* in 1865. He illustrates perfectly the theory of knowledge elaborated by Kant which replaced the ancient *theoria*.

Claude Bernard provides a detailed account of his discovery of 'the glycogenetic function of the liver' – the capacity of the liver to produce sugar. Bernard had observed, while carrying out tests, that there was sugar in the blood of the rabbits he dissected. He wondered about the origin of this sugar: did it come from ingested food or was it produced by the body, and, if so, which organ was responsible? He separated his rabbits into three groups: some were given food containing sugar; others were given food with no sugar; and the least fortunate were placed on a starvation diet. After several days, he analysed the blood of the rabbits, only to discover that, in every case, there was the same amount of sugar in their blood. This indicated that glucose did not derive from food, but was produced by the body.

The work of contemplation, the *theoria*, has changed completely since the Greeks: it is no longer a question of contemplation; science is no longer a spectacle but a job of work, an activity which consists of *making connections* between phenomena, in *associating* an effect (sugar) with a cause (the liver). And this is precisely what Kant, before Claude Bernard, had already formulated and analysed in the *Critique of Pure Reason*; namely the idea that science must define itself henceforth as a work of the associative faculty, or, to use his vocabulary, as a work of 'synthesis' – a word which in Greek means 'to put together', to 'combine'; just as an explanation in terms of cause and effect connects two phenomena: in this instance, sugar and the liver.

When I was young and I opened the *Critique of Pure Reason* for the first time, I was deeply disappointed. I had been told that he was perhaps the greatest philosopher of all time. Not only did I understand nothing, literally nothing, of what I read, nor could I understand why, from the opening pages of this mythical work, Kant was so preoccupied by a question which seemed to me trivial and uninteresting: 'How are synthetic *a priori* (not supported by fact; based on hypothesis or theory rather than experiment) judgments possible?' This does not at first seem an especially fertile subject for reflection – nor even, perhaps, at second sight.

For several years, I understood almost nothing of Kant. I was able to interpret the words and sentences, of course, and was able to give a plausible account of each concept, but the whole continued to make no sense. It was only when I realised the radically novel problem which Kant tried to address, in the wake of the collapse of inherited cosmologies, that I grasped the stakes raised by his opening question, which had previously struck me as purely 'technical'. In asking himself about our capacity to create 'syntheses', or 'synthetic judgements', Kant formulated the problem confronting modern science, that of scientific method: how does one devise laws which lay the ground for associations, that is, for coherent and revealing connections between phenomena whose ordinance or organising principle is no longer a given but must be introduced by us as an intervention, from outside.

A Revolution in the Moral Life

The theoretical revolution that Kant inaugurated was to have considerable consequences in the moral sphere. The new vision of the world forged by modern science had almost nothing in common with that of the Ancients. The universe as described by Newton, particularly, is no longer in any sense a place of peace and harmony; rather it is a world of blind forces and collision, in which we no longer know where to place ourselves, for the simple reason that it is infinite, without boundaries in space or time. As a result, it can no longer serve in any sense as a guide for morality. All of the questions of philosophy must therefore be completely reformulated.

We might say that modern thought puts mankind in the place of *cosmos* and divinity. It was on the rock of humanity that philosophers must build a new edifice of theory, of morality, and even new doctrines of salvation. It was up to man to introduce, by means of his intellectual labour, sense and coherence into a world which seemed no longer to possess meaning.

In terms of morality, you have only to consider the Declaration of Human Rights of 1789, the most visible and familiar external sign of a revolution without precedent in the history of ideas. It placed man at the centre of the world, whereas for the Greeks the world had been the centre of attention. Moreover, it not only made humans the sole beings on Earth worthy of full respect, but it proposed the equality of all humans, whether rich or poor, man or woman, white or black. In this case, modern philosophy is in the first place a *humanism*.

This transformation posed a significant question: if the ancient principles, cosmological and religious, had had their day, and men understood why this was the case, what new *theoria*, new ethics and new doctrine of salvation could take the place of those which had produced a *cosmos* and a divinity?

To answer this, modern philosophy placed at the centre of its cogitations a question which may seem very strange: what is the difference between men and animals? The philosophers of the seventeenth and eighteenth centuries were fascinated by the definition of what is an animal, believing that if they could establish the differences between man and beast, they could better discern man's 'specific difference', that which defines and is proper to him.

In the words of the great nineteenth-century historian, Jules Michelet, animals are 'our humbler brothers'. They are the beings closest to us; we can readily see how, from the moment that the idea of religion begins to falter and be replaced by the idea of man as centre of the world and subject of philosophical reflection, the question of what is 'proper to man' becomes intellectually crucial.

Modern philosophers now shared the notion not only that man had rights, but that he was the sole being to have rights – as the Declaration of 1789 affirms. If they now placed man above all other beings and assigned vastly increased importance to him, not merely over other animals, but also over a defunct *cosmos* and an increasingly doubtful divinity, then there must logically be something about man that distinguishes him from the rest of creation.

By starting with the debate about animals and, at the same time, the debate about man and the nature of humanity, we enter directly the key concepts of modern philosophy, and especially those of Jean-Jacques Rousseau, who, in the eighteenth century, would make the most decisive contribution.

The Difference between Animals and Humans According to Rousseau

If I were allowed to take only one text of modern philosophy to a desert island, then I would undoubtedly choose a passage from Rousseau's *Discourse on the Origin of Inequality*, published in 1755. We will come to the passage in a moment, but in order to understand it fully, you must first be aware that at the time of its writing there existed two classic criteria for distinguishing men from beasts: intelligence and sensibility (meaning affect, sociability, which included language).

In Aristotle, for example, man is defined as 'the rational animal', by which was meant a living being who possessed – as his 'specific' difference – an additional characteristic: the capacity for reason. For Descartes, the criterion of reason or intelligence was joined by a further property: that of affectivity (the emotion that lies behind action). He believed that animals are comparable to machines, or *automata* – machines that imitate the movement of a living creature, such as a clock – and it was an error to think of them as experiencing emotions. This would explain why they do not speak, even though they are equipped with organs which would make speech possible. They have nothing to express.

Rousseau proposed a radically new solution. His new definition of the human person was to prove inspirational, in that it enabled the founding of a new morality that was no longer 'cosmic' or religious but humanist – and 'a-theist'. For Rousseau, animals clearly possessed intelligence, sensibility, even the faculty of communication. Therefore it is not reason, or affectivity, or even language that differentiates the human being. On the contrary, everyone who has a dog knows perfectly well that the dog is more sociable and even more intelligent than, in some cases, certain human beings. In terms of sociability and intelligence, we barely differ from animals. Contemporary ethology – the scientific study of the function and evolution of animal behaviour – broadly confirms this diagnosis. We know today with certainty that there exists a highly developed animal intelligence and affectivity, which in the case of the great apes includes the acquisition of fairly sophisticated language-learning skills.

The true criteria for differentiating man and animal were to be sought elsewhere. Rousseau came to locate the difference in terms of man's liberty of action, what he called 'perfectibility' – broadly speaking, the capacity to improve oneself over the course of a lifetime; whereas the animal is guided from the outset by 'instinct' – is, in a manner of speaking, perfect 'from the start', from birth. It is clear that an animal is led by an unerring instinct, common to all members of his species, from which it can never really deviate. It is in this respect that the individual animal is deprived both of liberty and of the capacity to improve itself. It is 'programmed' by nature and, unlike man, cannot evolve further. Man, on the contrary, has the capacity to forge a personal history, whose progress is by definition open-ended and unlimited.

Rousseau expresses these ideas in a lucid passage, which should be read carefully before proceeding further:

> I see in every animal merely an ingenious machine to which nature has given senses to keep it going by itself and to protect itself, up to a certain point, from everything likely to distress or annihilate it. I see precisely the same things in the human machine, with the difference that nature alone brings everything to the activities of a beast whereas man contributes to his own, in his capacity as a free agent. The beast chooses or rejects by instinct, meaning that it cannot deviate from the rule prescribed for it, even when it might benefit from doing so, whereas man often deviates from such laws to his own detriment. This is why a pigeon would die of hunger next to a dish filled with choice meats and a cat next to a heap of fruit or grain, though either of them could get nourishment from the foods it disdains if only it had thought of trying them. This is why dissolute men give themselves over to the excesses that bring on fevers and death, because the mind perverts the senses and the will continues to speak when nature falls silent . . . Although the difficulties surrounding all these questions leave room for disagreement about this difference between man and beast, there is one further highly specific, distinctive and indisputable feature of man, namely his faculty for self-improvement – a faculty that, with the help of circumstances, successively develops all the others and that in man inheres as

> much in the species as in the individual; whereas an animal at the end of a few months has already become what it will remain for the rest of its life, and its species will be at the end of a thousand years what it was in the first year of that millennium. Why is only man prone to turn senile? Is it not the case that he thus returns to his primitive state and that, while the beast that has acquired nothing and hence has nothing to lose is always left with its instincts, man, losing through age or some accident all that his 'perfectibility' has enabled him to acquire, ends by sinking lower than the beast? the origin of (*Discourse on the Origin of Inequality*)

What exactly is Rousseau saying? Let us begin with the example of the cat and the pigeon. He is saying that animals operate within a framework of invisible codes, a kind of 'software' from which they are unable to escape. It is as if the pigeon is the prisoner of his programming because it can only eat grain, and the cat likewise because it is a carnivore. There is little possibility, if any, for them to depart from these scripts. No doubt a pigeon could ingest a very small quantity of meat, and the cat could nibble at a few blades of grass, but, all in all, their natural programme leaves them almost no room for manoeuvre.

The human condition is very different, because he is capable of change. In fact, he can deviate from all the rules prescribed for animals. For example, he can commit *excess*: drink and smoke, to the point of killing himself, which is impossible in nature. Or, as Rousseau says in a formula which announces the whole of modern politics, in man 'the will continues to speak when nature falls silent'. In the case of animals, nature speaks continually and forcibly, so much so, that the animal can do nothing but obey this voice. In man, it is a certain *lack* of determination that speaks loudest: although nature does, of course speak, as we are constantly reminded by biologists: we too have bodies, genetic programming, the rule of DNA, of the genome transmitted by our parents. Nevertheless, man can disregard these natural rules and even create a culture which opposes them virtually point by point – as we see in the culture of democracy which tried to thwart the logic of natural selection so as to secure the safety of the weakest.

One example of the transcendence of human will over natural programming and its capacity for deviation or excess is far more striking: the phenomenon of evil. It powerfully confirms Rousseau's argument about the anti-natural and therefore non-animal character of human will. It is as if only mankind is capable of behaving in what might be termed a *diabolical* manner.

But are animals not as capable of aggression and cruelty as man? At first sight, this might appear true, and one could give several examples which animal lovers usually prefer not to discuss. When I was a child, living in the country, there were dozens of cats, which I would regularly see destroying their prey with what seemed great cruelty: eating mice alive, toying for hours with a bird whose wings they had broken . . .

But *radical* evil, which in Rousseau's perspective is unknown to the animal order and a specifically human invention, is to be found elsewhere: it consists not simply of 'doing ill', but of *adopting evil as its project*, which is a quite different proposition. The cat mistreats the mouse, but this is not the purpose of its natural instinct to hunt mice. On the other hand, everything suggests that the human being is capable of consciously organising himself so as to inflict the greatest possible evil upon his neighbour – what traditional theology designates *malice*: the evil spirit within us.

This malice, unfortunately, seems to be unique to man. There is an absence in the animal world, in the natural order, of anything resembling the phenomenon of torture. Today one can visit in Ghent, in Belgium, a mind-boggling museum: a museum of torture. Here, exhibited in glass cases, are the appalling products of the human imagination: chisels, knives, pliers, head-clamps, instruments for pulling nails and crushing fingers, and a thousand other devices.

Admittedly, animals frequently eat each other alive, which strikes us as cruel, but evil as such is not their intention, and their apparent cruelty stems from their indifference to the suffering of others. Even when they appear to kill 'for pleasure', they are only exercising their instinct. Anyone who has owned cats will have seen them 'torturing' their prey, but it is because in doing so they exercise and perfect their hunting skills. What seems like cruelty is linked to the relations between predator and prey.

But the human is not subject to a rule of indifference. He commits evil knowingly and, on occasion, enjoys it. Contrary to the animal, the human can make a conscious choice to do evil. Everything would seem to indicate that torture goes beyond the logic of any situation. Some would argue that sadism is a pleasure like any other, and that it must be encoded somewhere in human nature. But this explanation explains nothing. It is intentionally deceptive: as if sadism can be justified by evoking the pleasure taken in the suffering of another. The real question is the following: why is there so much *gratuitous* pleasure in transgression, even when it serves no end?

Man tortures man for no reason, other than torture itself. Why did a Serb militia (as noted in one report on war crimes committed in the Balkans) force a Croat grandfather to eat the liver of his still living grandson? Why, for that matter, do some cooks happily dismember live frogs and eels, when it would be simpler and more logical to kill them first? The fact is that humans take it out on animals when the human material is in short supply, but never on automata that do not suffer: have you ever seen a man take pleasure in torturing a watch or a pendulum clock? I do not believe there is a convincing 'natural' explanation for committing evil; it seems to belong to another order than that of nature. It serves no purpose, and is in most cases counter-productive.

It is this anti-natural vocation, this constant possibility of excess that we see in the human eye: because it reflects not only nature, it can seem to express the worst; but equally, and for the same reason, the best: absolute evil or an astonishing capacity for selflessness. It is this principle of excess that Rousseau refers to as liberty.

Three Consequences of the New Distinction between Man and Beast

The consequences of this redefinition of man are enormous, and three aspects were to have considerable consequences for the ethical and political spheres.

First: humans, contrary to beasts, become invested with what might be called a *double historicity*: there is the history of the individual, which is referred to as *education*; and there is the history of the human species, or human societies, their *culture and politics*. When we consider animals, the case is quite different. From Antiquity onwards we have descriptions of 'animal societies', a perfect example being a bee colony. From these we can deduce that their behaviour has been the same, *exactly* the same, for thousands of years: their habitat has not changed and neither has their method of gathering pollen, nourishing the queen and dividing responsibilities. Human societies change incessantly: if we could travel back in time two hundred years we would not recognise Paris or London or New York. But we would have no difficulty in recognising an anthill or a cat chasing a mouse.

You might wonder about animals learning to hunt with their parents. Is that not a form of education? Yes, but it is a short-term 'apprenticeship' that stops as soon as the objective has been attained; a human education is unlimited, ending only with death.

Some animals don't even serve an apprenticeship; they behave as miniature adults from the moment of their birth. In the case of young marine turtles, as soon as they emerge from the egg they know instinctively how to find their way unaided to the ocean. They are immediately able to walk, swim and eat – everything they need to survive. Whereas my children have remained happily at home until their twenties!

So, Rousseau touched on an important issue when he spoke of liberty and perfectibility. How can we explain this difference between the young turtle and the human infant if we do not accept some form of liberty. The baby turtle possesses neither a personal history (an education) nor a political and cultural history, and is from its birth and for always driven by the regime of nature, by instinct. The human individual can evolve indefinitely, educate himself 'for life', and embark on a personal history of which nobody can say how or when it will end.

Second: Jean-Paul Sartre said that if man is free, there is therefore no 'human nature', no human 'essence', no definition of humanity which precedes and determines his individual existence. In a little book, *Existentialism is a Humanism*, which I would advise everyone to read, Sartre develops this idea by asserting that, in the case of man, 'existence precedes essence'. This is pure Rousseau, almost word for word. Animals have an 'essence', common to their entire species, which precedes their existence as individuals: there is a cat essence, an essence of pigeon – and this instinct or 'essence' is common to the entire species, so much so that the individual identity and existence of each individual is wholly determined by it: no cat, no pigeon, can swerve away from this essence which suppresses all individual action.

But, with humans, the opposite is true: no essence predetermines it, no programme can ever succeed in entirely hemming it in; no system can imprison it so absolutely that it cannot emancipate itself. I am born *into* a society: as a man or woman, native or immigrant, rich or poor, aristocrat or labourer, but these initial categories do not define me for the rest of my life. I can be a woman and decide not to have children; I can be born poor and underprivileged, but become rich; I can be born French, but adopt another language and change nationality, and so on.

From this notion that there is no human nature, that man's existence precedes his essence – according to Sartre – we arrive at an unanswerable debate about racism and sexism. What is racism? What is sexism? It is the idea that there exists an essence exclusive to each race, or to each sex, and that individual members are wholly contained by it. The racist says that 'the African is child-like', 'the Jew is intelligent', 'the Arab is lazy'; and from this use of the definite article – 'the' – we know that we are dealing with a racist, someone who believes that all individuals of the same group share the same traits, or 'essence'. Likewise for the sexist, who believes that it is part of the 'natural' essence of a woman to be more emotional than intelligent, more kind-hearted than courageous, whose duty it is to have children and stay in the kitchen.

Rousseau destroyed this type of reasoning at the root. The human individual is free: endlessly improvable, and in no sense programmed by characteristics supposedly linked to race or gender. Of course, the individual is born into a particular 'situation', but that is not equivalent to a software programme with no margin for manoeuvre. It is this margin, this gap, which is the distinctive property of mankind, and which racism – 'inhuman' as it is – would destroy at all cost.

Third: because he is free, because he is not imprisoned by any natural *code* or historical determinant, the human is a moral being who can choose freely to act in a good way or an evil way. Who would dream of accusing a shark who has just eaten a surfer of acting badly? When a lorry causes an accident, it is the driver whom we judge, not the vehicle. Neither the fish nor the vehicle are responsible for the effects, however harmful, they have upon a human being.

From Rousseau's perspective one must distance oneself from the real in order to judge well or poorly, just as one must distance oneself from appearances – natural or historical – to acquire what is termed the 'critical spirit', without which no value judgement is possible. Kant once said of Rousseau that he was 'the Newton of the moral universe', in that man is unendingly torn between egotism and altruism, just as the Newtonian universe is pulled between centripetal and centrifugal forces. He meant primarily that with his invention of human freedom, Rousseau was to modern ethics what Newton had been to the new physics: a pioneer, a founding father without whom we would never have been able to free ourselves from ancient ideas regarding the *cosmos* and the divine principle. By identifying the principle of a differentiation between animal and human, Rousseau finally made possible a new moral vision of the world.

The Heritage of Rousseau: Man as a 'Denatured Animal'

In the 1953 novel *You Shall Know Them*, by Vercors (the pseudonym of Jean Bruller, who worked in the French Resistance), there is an interesting interpretation of these notions of Rousseau. Here is a brief summary of the plot: in the 1950s, a team of British scientists sets out for New Guinea, in search of the celebrated 'missing link' – the primate that links man and animal. They hope to discover some unknown great ape fossil but by sheer chance they come upon an entire colony of evolutionary 'intermediates', which they call 'Tropis'. These are quadrupeds – monkeys – but they live, like the troglodytes, in caves, and, most importantly, they bury their dead. This perplexes our explorers because it does not resemble any known animal custom, and the Tropis seem to speak a rudimentary language.

Where do they sit on the imaginary ladder between the human and the animal? The answer to this question becomes urgent when an unscrupulous businessman plans to domesticate the Tropis in order to make slaves of them. If they are animals he can get away with it, but if they were classified as human it would be unacceptable, as well as illegal. How can the question be resolved?

The hero of the novel makes the necessary sacrifice: he gets a female Tropi pregnant, the act itself proving that this species is closer to the human species (biologists agree that, with rare exceptions, only members of the same species can reproduce). Is the child to be classified as human or animal? A decision must be taken, for the father has decided to kill his child, in an attempt to force the law to make a decision. A legal case gets underway, which grips the entire nation, and very soon involves the world's press. The most eminent specialists are summoned to appear: anthropologists, biologists, palaeontologists, philosophers, theologians . . . All of them disagree with each other, but their respective arguments are so persuasive that none succeeds in winning over the others.

At last, the judge's wife comes up with the decisive criterion: if they bury their dead, she argues, the Tropis must be human. And the reason is that this ritual indicates a metaphysical awareness of reality. As she expresses it to her husband: 'In order to ask a question, one must be two: the one who asks, and the one who is asked. Because he is part of nature, the animal cannot question nature. It seems to me that this is the distinction we are looking for. The animal is one with nature; whereas man and nature make two.' This is a perfect translation of Rousseau's insight: the animal is a natural creature, and merged completely with nature; man on the contrary is beyond nature.

This needs further refinement. Why, in these circumstances should the criterion of distance from nature be more important than any other? After all, animals do not wear watches or carry umbrellas, do not drive cars or listen to MP3s, do not smoke cigarettes or drink wine. There can be no doubting the answer: distance from nature is the only criterion that counts decisively on both an *ethical and cultural* plane: this distance makes it possible for us to engage with the history of culture, rather than remaining the hostages of nature. This distance enables us to interrogate reality, to judge and transform the world, to invent 'ideals', to distinguish *between good and evil*. Without it, no morality would be possible. If nature were our code, no ethical decision could ever have occurred – the Tropis would not even consider burying their dead. While human beings concern themselves with the fate of animals, trying to save the whales, for example, have you ever known a whale show any interest in the fate of a human being, except in fairy tales?

With this new 'anthropology', this new definition of what is peculiar to man, Rousseau prepared the way for modern philosophy. And from here the dominant secular morality of the past two hundred years came into being: namely, the ethical system of the greatest German philosopher of the eighteenth century, Immanuel Kant, the repercussions of which were to have considerable impact within the French republican tradition.

Kantian Ethics and the Foundations of the Republican Ideal

It was Kant who illustrated the two most striking consequences, for morality, of this new definition of man as a free agent: first, the idea that moral virtue resides in actions that are disinterested and not for private or selfish gain; and second, that these are directed towards the common and 'universal' good. These are the two principal pillars – *disinterestedness* and *universality* – of the ethics Kant was to set forth in his famous *Critique of Practical Reason* (1788). They were to become so generally accepted – especially by the French republicans – that they came to define what might be termed *the* modern morality.

The truly moral – truly 'human' – action (and it is

significant that the two terms begin to overlap) becomes first and foremost the disinterested action: in other words, man's capacity to act independently of his natural urges, which leads him inexorably in the direction of egotism. The decision to resist egotistical temptations is described by Kant as 'Good Will': although I am inclined (since I am *also* an animal) to satisfy my personal interests, I am equally able to ignore them and to act disinterestedly and altruistically.

What is perhaps most striking about this new moral perspective – both anti-naturalist and anti-aristocratic (since, contrary to the distribution of natural talents, the talent for freedom is supposedly innate in all of us), is that the ethical value of disinterestedness seems self-evident, so that we do not even pause to reflect upon it. If I discover, for example, that somebody who appears kindly and generous towards me is doing so in the hope of obtaining some advantage or other which they conceal from me (the hope of inheriting my fortune, perhaps), the presumed moral value attributed to their actions immediately disappears. In the same sense, I attribute no especial moral value to the taxi-driver who agrees to take me somewhere, because I know that he does it for money. On the other hand, I would be very grateful to the cabbie who picks me up, without any motive of self-interest, when I am hitching to work on a day of transport strikes.

These examples all point towards the same conclusion for the new Humanist: that virtue and disinterested action are inseparable. We must be capable of acting freely, without being programmed by any natural or historical codes, to attain to the sphere of disinterestedness and unscripted generosity towards others.

The second fundamental ethical conclusion to be derived from the thought of Rousseau and his disciples is linked directly to the first, and concerns the ideal of the common good, on the universality of moral actions transcending individual private interests. The good is no longer linked solely to my own personal interest, or that of my family or tribe. It does not exclude these, of course, but it must also accommodate the interests of others, even of humanity as a whole.

Here, too, the link with freedom is explicit: nature is by definition specific and particular; I am a man or a woman; I have a particular body with its own tastes, desires and inclinations which are not necessarily altruistic. Were I always to follow my animal nature, it is likely that the common good and the general interest would have to wait a long time before I paid them any attention. But if I am free, I can resist my animal nature, even in small doses, because I distance myself from myself, in a way – I can draw closer to others, enter into solidarity with them, and take account of their requirements. Which is, surely, the minimum condition for a mutually considerate and harmonious shared life.

Freedom, the virtue of disinterested action ('good will'), and concern for the general welfare: these are the three key concepts which define the modern morality of duty, and which Kant was to express in the form of absolute commandments, known as *categorical imperatives*. Given that it is no longer a question of imitating nature, of taking nature as our guide, but actually resisting nature (and our innate egotism), it becomes clear that the achievement of the good, of the general interest, does not happen of its own accord, but that on the contrary it encounters resistance.

If we were naturally inclined towards the good, there would be no need for absolute commandments. Most of the time we have no difficulty in recognising what we ought to do in order to act well, yet we constantly make exceptions, for the simple reason that we place ourselves before others. This is why the categorical imperative invites us to 'make an effort', to constantly try to improve ourselves.

Within this new morality is the notion of merit: we all have difficulty in doing our duty, in following the commands of morality, even as we recognise their validity. There is therefore merit in acting well, in preferring the general interest to our private interest, the common good to egoism. The reason for this is simple: while we are unequal in terms of our innate talents – strength, intelligence, beauty and so on – in terms of merit we are all equal. For, in Kant's perspective, it is merely a question of motive, of 'good will'. And this is the property of all of us, strong or weak, beautiful or ugly.

Aristocratic and Meritocratic Models

To understand the novelty of this modern ethos, we need to see what is new about the meritocratic model of virtue, as opposed to older and aristocratic versions. In 1755 in Lisbon an earthquake wiped out several thousand people in a matter of hours. It had a devastating effect across Europe, and many philosophers questioned the meaning of natural catastrophes: should it be this version of nature, hostile and malevolent, that we should take for our guide, as instructed by the Ancients? Not only does nature no longer seem remotely good, most of the time men find themselves having to oppose the natural order to arrive at any notion of good. And this is as much the case *within ourselves* as around us.

If I listen to my inner nature, it is an uninterrupted and insistent babble of egotism that speaks, urging me to follow my private interests to the detriment of others. How could I for a moment imagine connecting with the common good if I content myself with listening exclusively to the demands of my own nature? With one's inner nature others can always wait.

The crucial question confronting ethics in modern life is how do we remake a coherent world between humans without involving nature – which is no longer a *cosmos* – or a divinity, which has no meaning other than for its followers?

The answer, which defines modern humanism ethically as much as politically or juridically, is as follows: such a world must be founded solely upon the will of men, provided that they agree to restrain themselves, to limit themselves by acknowledging that their individual freedom must sometimes stop where that of others begins. It was only from this voluntary restriction of our desire for expansion and conquest that a peaceful and reciprocal relation could be created between men – 'a new *cosmos*', one might say, but this time ideal rather than natural, and socially constructed rather than pre-existing. Kant would designate this 'second nature', this fiction of coherence devised by the free will of humans in the name of collective values, 'the kingdom of ends'. In this 'brave new world', the world of the will rather than of nature, humans would be treated finally as 'ends' rather than 'means': as beings possessed of dignity who were not raw material in the service of supposedly higher objectives. In the ancient world, the human individual was no more than an atom among atoms, a fragment of a far superior reality. Now, the individual was the centre of the universe, and the creature beyond all else entitled to absolute respect.

To fully understand how revolutionary Kantian morality was, it is worth comparing it to classical ethics, specifically the notion of 'virtue'. Ancient commentators characteristically defined virtue or excellence as an extension of nature; as a realisation, more or less perfect, for each being, of what constitutes its nature and thereby indicates its 'function' or its purpose. In this given nature of every being could be read its final destiny. Which is why Aristotle begins the *Nicomachean Ethics* by reflecting on what is the final purpose of man among other creatures: 'For just as for a flute-player, a sculptor, or any artist, and, in general, for all things that have a function or activity, the good and the "achieved" is thought to reside in their function, so it would seem to be for man, if he has a function.' This being so, it would be absurd to suppose 'that a carpenter or a tanner have certain functions and activities, but that man has none, and is born without a function.' (1097 b 25)

Here we are in the presence of a nature that fixes the purposes of man and gives a direction to ethics.

The philosopher Hans Jonas noted that, in the ancient idea of the cosmos, ends are 'domiciled in nature', hard-wired in nature. Which does not mean that in the accomplishing of his specific task, the individual does not encounter difficulties, that he does not need to exercise will and the faculties of judgement. But it remains the case, for moral as for all other activities – learning to play a musical instrument, for example – that practice may make better, but above all else *talent* makes perfect.

While the aristocratic order did not completely exclude a notion of the individual will, only natural gifts could indicate the way forwards and remove the obstacles with which it was strewn. This was why 'virtue' (or 'excellence', the terms here mean the same) was defined as a 'just measure', a middle way between extremes. In terms of fully realising our natural destiny, it was clear that this could only be found in an intermediate position: for example, courage was to be found equidistant from cowardice and recklessness, just as good eyesight lay between near-sightedness and long-sightedness. The just measure had nothing to do with taking a 'centrist' or moderate position; it was a search for perfection. In this sense it could be said that an individual realises to perfection its nature or essence when it is equally remote from poles which, because they are at the limit of their definition, verge upon monstrosity. The monstrous is that which, by its 'extremism', ends by distorting its proper nature, an unseeing eye, for example, or a three-legged horse.

Early on in my philosophy studies I had great difficulty in understanding how Aristotle could speak seriously of a horse as having a 'virtuous' eye. The text in question, from the *Nicomachean Ethics*, perplexed me:

> We may remark, then, that every virtue or excellence both brings into good condition the thing of which it is the excellence, and makes the work of that thing to be done well; e.g. the excellence of the eye makes both the eye and its works good; for it is by the excellence of the eye that we see well. Similarly the excellence of the horse makes a horse both good in itself and good at running and at carrying its rider and awaiting the attack of the enemy.' (1106 a 15)

I could not see what the notion of 'virtue' contributed to this case. In an aristocratic perspective, however, such a proposition held no mystery: the 'virtuous' individual is not one who attains a certain excellence thanks to openly acknowledged effort, but one who functions well, even excellently, according to the nature and purposes which are innately his. And this principle applies to things and animals as much as to humans whose happiness is linked to this accomplishing of self.

At the heart of such an ethical vision, the question of limits thus receives its 'objective' solution: it is in the order of things, in the reality of the *cosmos*, that we must seek instruction; just as a physiologist seeking to understand the function of organs and limbs simultaneously recognises the limits within which each exercises its function. Just as we would not exchange a liver for a kidney, without injury, each of us in society must find his proper place and confine himself to it; otherwise the law will intervene to restore order and harmony, and render – in the famous formula of Roman law – 'to each his own'.

The difficulty, for us nowadays, is that such a 'cosmic' reading became impossible; for lack of a *cosmos* to interpret and for lack of a nature to decipher. One could therefore describe the cardinal distinction between the cosmological ethics of the Ancients and the meritocratic, individualist ethics of republican Moderns, beginning with the anthropology of Rousseau, as follows: for the Ancients, *virtue*, understood as excellence of its kind, is not opposed to nature; it is none other than the successful fulfilment of an individual's natural aptitudes. For the philosophies of human freedom, most notably for Kant, virtue takes a different form, as the struggle for release from what is natural within ourselves.

Our nature inclines us naturally to egotism, and if I wish to give a place to others, if I wish to attune my freedom to being in accord with the freedom of others, I must make an effort – or restrain myself – and it is on these terms alone that a new order of peaceful co-existence between individuals becomes possible. Here is the future of virtue, no longer in the fulfilment of a well-endowed natural self. It is through the exercise of a new virtue alone that a new *cosmos*, a new order of things, becomes possible; one founded on man and not on a pre-existing *cosmos* or a divinity.

On the political plane, this new order of things was to display three characteristic features, directly opposed to the aristocratic world of the Ancients: categorical equality of status, individualism and the assignation of value to the idea of work.

If we identify virtue with natural endowments, all

individuals are not equal. From this perspective, it is logical to create an aristocratic order, in other words an unequal world; not just a natural hierarchy of beings, but one in which the best are on top and the rest below. But, if we locate virtue in freedom – rather than in nature – all individuals are equal, and democracy becomes inevitable.

Individualism is a consequence of this reasoning. For the Ancients, the *cosmos* is infinitely more significant than its constituent parts. This can be described as 'holism', deriving from the Greek *holos* ('all', 'everything'). For the Moderns, there is no longer anything sacred about the All, since there is no longer any divinely ordained and harmonious *cosmos* within which we must find our place. Only the individual counts: we no longer have the right to sacrifice the individual in order to maintain the universal (the All), for the latter is no longer anything other than an aggregate of individuals, within which each human being remains 'an end in himself'.

So, the term individualism is far from meaning egotism, as is commonly thought; on the contrary, it is the birth of a moral sphere within which individuals – persons – are valued by their capacity to break free of the logic of their natural egoism, in order to construct a man-made ethical universe.

Finally, in the same perspective, work becomes the defining activity of man: a human being who does not work is not merely poor – without income – but impoverished, in that he cannot realise his potential and his purpose on Earth. His aim is to create himself by remaking the world, to transform it into a better place by the sheer force of his 'good will'. In the aristocratic world-view, work was considered to be a defect, a servile activity – literally, reserved for slaves. In the modern world-view, it becomes an arena for self-realisation, a means not merely of educating oneself but also of fulfilment and improvement.

We have seen how modern science exploded the very idea of a *cosmos* and obedience to divine injunctions, and how ancient systems of morality ran into difficulties. The new definition of man proposed by modern humanism – notably by Rousseau – prepared the way for the birth of a new morality, beginning with that of Kant and French republicanism.

I began my account of modern philosophy with Rousseau and Kant – eighteenth-century philosophers – whereas the true rupture with antiquity occurred in the seventeenth century, specifically with Descartes. Descartes is the true founder of modern philosophy, and it is important that we have some idea of the reasons why he represents both a point of rupture and a point of departure.

The Origin of Modern Philosophy

Cogito ergo sum – 'I think therefore I am' – is one of the most universally celebrated and significant of philosophical utterances, and rightly so, since it draws a line in the history of Western thought, and inaugurates a new epoch; that of modern humanism, at the centre of which is what we shall refer to as 'subjectivity'. What exactly do we mean by subjectivity?

At the start of this chapter, we saw how the verses

of John Donne ("Tis all in pieces, all coherence gone') epitomised the state of mind of an age of uncertainty, in which everything had to be reconstructed: a theory of knowledge, a new ethics and, perhaps most of all, a doctrine of salvation. For this, a new first principle was required, which was neither *cosmos* nor divinity. This was to be none other than man, or, as the philosophers would say, the 'subject'.

It was Descartes who 'invented' this new first principle, prior to its application by Rousseau and Kant. It was Descartes who transformed the doubt linked to the disappearance of ancient certainties into a formidable tool for reconstructing from scratch the entire edifice of philosophical thought. In his two fundamental works, *The Discourse on Method* (1637) and the *Meditations* (1641), Descartes conceives, under various guises, a form of philosophical fiction (or 'method', as he terms it). He forces himself to call into question each and every one of his ideas, without exception, even the most settled and self-evident truths; such as, for example, the existence of objects outwith myself, that I am seated on a chair etc. In order to be certain about doubting all certainties without exception, he even imagines the hypothesis of an 'evil genie' who amuses himself by deceiving Descartes about absolutely everything.

Descartes adopts an attitude of total scepticism, taking nothing on trust . . . except that, at the end of the day, there does remain a certitude which resists everything and vigorously stands its ground, a conviction that holds good under even the most extreme doubt. And this is the certitude that, according to which if I am thinking these things, even in a state of uncertainty, I myself must therefore be something that exists! It may well be the case that I am continually making errors, that all my ideas are false, that I am permanently deceived by an evil genie – but in order for me to be deceived, or to deceive myself, I must at the very least be something that exists! A conviction remains that is resistant to all doubt, however general, and it is the certainty of my own existence. From which comes the formula with which Descartes concludes his investigations: 'I think therefore I am'.

The experience of radical doubt which Descartes depicts – and which may strike you initially as outlandish – offers three new ideas, which make their appearance for the first time in the history of thought. These three ideas were destined for a remarkable posterity and they are of fundamental importance to modern philosophy.

First: each time Descartes stages a new drama of doubt, it is not merely an intellectual game; it aims to arrive at a new definition of truth. By examining close up, and with scrupulous care, the only certainty which categorically resists every challenge – the *cogito*, in effect – he will arrive eventually at a reliable truth-criterion. We can even say that this method of reasoning will lead to a definition of truth as that which resists doubt, as that of which the individual subject can have absolute certainty. Thus a state of subjective consciousness – certainty – becomes nothing less than the new criterion of truth. And this will give you an idea of how central the category of *subjectivity* becomes for the project of Modernity.

It is henceforth exclusively in terms of the 'subject' that the surest measure of truth is to be found (whereas the Ancients defined truth in *objective* terms; for example, when I say it is night, this proposition is true if and only if it corresponds to objective reality, to the facts themselves, whether I am certain of myself or not). Of course, the subjective criterion of certainty was not unknown to the Ancients – it is discussed in the dialogues of Plato – but with Descartes it was to take on a primordial authority and override all other criteria.

Second: even more decisively, in political and historical terms, was the idea of the 'tabula rasa' – the absolute rejection of all preconceptions and all inherited beliefs deriving from tradition. By putting in doubt all received ideas, without distinction, Descartes at a stroke invented the modern notion of *revolution*. As the nineteenth-century political thinker and historian Alexis de Tocqueville was to remark, the men who started the French Revolution of 1789, and who we refer to as the 'Jacobins', were in fact 'Cartesians' who had left school and taken to the streets.

One could say that the revolutionaries repeated in the historical and political sphere what Descartes had initiated in the sphere of abstract thought. The latter declared that all past beliefs, all ideas inherited from family or state, or indoctrinated from infancy onwards by 'authorities' (masters, priests) must be cast in doubt, and examined in complete freedom by the individual subject. He alone is capable of deciding between true and false. In the same way, the French revolutionaries declared that we must cast aside all the paraphernalia of the Ancien

Regime; as one of them, Rabaut Saint-Etienne, declared, in a wholly 'Cartesian' maxim which came to stand out as a milestone, the Revolution could be encapsulated in a single sentence: 'Our history is not our destiny.'

Just because we have lived since time immemorial in a regime that is an aristocracy and a monarchy, with established privileges and inequalities, we are not for ever obliged to continue doing so. Nothing compels us to continue to observe traditions for ever. On the contrary, when they are not good, we must reject them and change them. We must know how to 'make a blank slate of our past' in order to create from scratch – just as Descartes, having cast all prior beliefs in doubt, undertook the total reconstruction of philosophy upon a solid basis: namely an immovable certitude, that of a subject who takes responsibility for himself, and who trusts henceforth to himself alone.

In both cases – with Descartes as with the revolutionaries of 1789 – the human subject becomes the foundation of all thought, and the agent of all change: through the decisive experiment of the *cogito*, the democratic and egalitarian abolition of the privileges of the Ancien Regime and the (entirely unprecedented) declaration of the equality of all men.

Note that there is a direct link between the two ideas above; between the definition of truth as certitude of the subject and the founding of a revolutionary ideology. If we must make a *tabula rasa* of the past and subject to the most rigorous process of doubt all those opinions, beliefs and preconceptions which have not undergone minute examination, this is because it is proper to believe, to 'admit to credence' (in Descartes' words) only that of which we can be absolutely certain in our own minds. From which also proceeds a new version of nature, founded on individual conscience rather than tradition, of a unique certitude which compels recognition before all other kinds: that of the individual subject in his relation to himself. Thus it is no longer belief or faith which enables us to reach an ultimate (Christian) truth, but awareness of self.

Third, an idea whose unprecedented revolutionary power in the age of Descartes is hard for us now to imagine: whereby we must reject all 'arguments from authority'. The expression 'arguments from authority' means all beliefs with a claim to absolute truth imposed externally by institutions endowed with powers that we have no right to dispute: family, schoolmasters, priests and so on. For example, if the Church decrees that the Earth is not round and does not travel around the Sun, you must do likewise, and if you refuse, you run a high risk of ending up burned at the stake or being compelled to confess publicly that you are in the wrong, like Galileo, even if you are entirely in the right.

It is these arguments from authority that Descartes abolished, with his radical doubt. With his invention of the 'critical spirit', freedom of thought, he is the rightful founder of modern philosophy. The idea that one must accept an opinion because it is maintained by external authority, of whatever kind, became so repugnant to the Modern spirit as virtually to define Modernity. It is true, we sometimes extend our trust to a person or an institution, but the gesture has ceased to have any of its traditional meaning: if I agree to follow another's judgement, it is because I have formulated good reasons for doing so, not because this other imposes its authority externally without my assent.

I hope that you can now grasp a little more clearly how one can say that modern philosophy is a philosophy of 'the subject' – a humanism – with man at the centre of everything. And, finally, a few words about the new doctrines of salvation: in the absence of a *cosmos* or a God, according to strict humanism principles, the idea of salvation would seem virtually unthinkable. It is difficult to see where any notion of salvation might rest, in order to circumvent the fear of death. So difficult, in fact, that, for many people, the question of salvation was to vanish completely. Or it became confused with the question of ethics. This confusion is so frequent, even today, that I shall attempt to dispel it before embarking on the modern responses to the ancient question of salvation.

From Ethical Questions to the Question of Salvation

To reduce to their essentials these new ideas, we could define secular morality as an ensemble of values, expressed by obligations and imperatives, which ask us to pay a minimum of respect to others without which a shared and peaceful coexistence becomes impossible. What our societies – which make an ideal of the rights of man – ask us to respect in others is their dignity as our equals, their right to freedom, notably freedom of opinion, and their right to wellbeing. This is described in the famous maxim, 'My freedom stops where that of another begins'.

No one can doubt that moral laws should be indispensable and rigorously applied, for in their absence a war of everyone against everyone else begins to loom on the horizon. Such laws appear therefore to be the necessary condition of that peaceful coexistence which favours the emergence of a democratic order. They are not sufficient in themselves, however, and I aim to convince you that ethical principles, however precious they may be, have no purchase whatsoever on the great existential questions that were formerly taken care of by the doctrines of salvation.

I would like you to imagine that you own a magic wand which allows you to arrange matters so that everyone in the world today begins to observe to the letter the ideal of respect for others embodied in humanist principles. Suppose that, everywhere in the world, the rights of man were scrupulously observed, with everyone paying respect to the dignity of everyone else and the equal right of each individual to partake of those famous fundamental rights of freedom and happiness. We can hardly begin to comprehend the unprecedented revolution that such an attitude would introduce into our lives and customs. There would be no wars or massacres, no genocide or crimes against humanity. There would be an end to racism and xenophobia, to rape and theft, to domination and social exclusion, and the institutions of control or punishment – police, army, courts, prisons – would effectively disappear. So, morality counts for something, and this exercise suggests the degree to which it is essential to our common life; and, at the same time, how far we actually are from its realisation.

Yet, such a miracle would not prevent us from getting old, from looking on helplessly as wrinkles and grey hairs appear, from falling ill, from experiencing painful separations, from knowing that we are going to die and watching those we love die. In the end, nothing will save us from getting bored and finding that everyday life lacks zest. Even were we saints, immaculate apostles of the rights of man and the republican ethos, nothing would guarantee us a fulfilled emotional life. Literature teems with examples of how the logic of morality and that of love obey contradictory principles. Good morals have never saved anyone from being deceived or abandoned. Unless I am much mistaken, none of the love stories recounted in the classic novels proceed from humanitarian motives. If the implementation of the rights of man makes possible a peaceful coexistence, these rights do not of themselves give meaning or purpose or direction to human existence.

This is why, in modern life as in the ancient world, it was necessary to devise something – beyond morality – to take the place of a doctrine of salvation. The difficulty is that, in the absence of a *cosmos* or a God, it becomes especially difficult to think this through. How do we confront the fragility and finiteness of human existence, the mortality of all things in this world, in the absence of any principle external to and higher than humanity? This is the problem which the modern doctrines of salvation have tried to solve – for better or worse – and, it has to be admitted, usually for the worse.

The Emergence of Modern Spirituality

To reach this point, the Moderns turned in two main directions. The first – I will not hide the fact that I have always found it faintly ridiculous, but it has acquired such predominance over two centuries that we cannot ignore it – are what we might call the 'religions of earthly salvation', notably scientism, patriotism and communism. Unable to continue believing in God, the Moderns invented substitute-religions, godless spiritualities or, to be blunt, ideologies which, while usually professing a radical atheism, cling to notions of giving meaning to human existence, or at least justifying why we should die for them. From the scientism of Jules Verne to the communism of Marx, passing via the nineteenth century's brand of patriotism, these grand human – all too human – utopias have all at least shared the merit (albeit a doomed merit) of attempting the impossible: resuscitating great notions without stepping outside the frame of humanity – as the Greeks did with their *cosmos* or the Christians with their God. Here are three ways of saving one's life, or justifying one's death, which come to the same thing, by sacrificing it for a nobler cause: whether that means the revolution, the homeland or the truths of science.

With these three 'idols', as Nietzsche would term them, the essentials of faith were rescued: to consecrate and if necessary sacrifice one's life to an ideal was to preserve the conviction of being 'saved'. To give a grotesque example, I will quote a low point from the history of the French press. It concerns an article in *France nouvelle*, the weekly organ of the communist party, published on the morning after the death of Stalin. Stalin was then the head of the Soviet Union, the pope of world communism, so to speak, and considered by the faithful as a hero, despite his crimes.

On 14 March 1953 the French Communist Party composed the newspaper's front page in terms that seem today incredible, but which capture perfectly the abidingly religious idea of death at the heart of a doctrine which nevertheless saw itself as radically materialist and atheist. Here is the text:

> The heart of Stalin, illustrious comrade-in-arms and renowned successor of Lenin, the chief, the friend, the brother of workers everywhere and in all countries, has ceased beating. But Stalinism lives on, and is immortal. The sublime name of the inspired master of world communism will shine with blazing clarity across the centuries and shall for ever be pronounced with love by a grateful humanity. To Stalin we shall remain faithful for evermore. Communists everywhere will endeavour to deserve, by their untiring devotion to the sacred cause of the working class . . . the honorary title of Stalinists. Eternal glory to the great Stalin, whose masterly and imperishable scientific works shall help us to rally the majority of humanity. (*France nouvelle*, 14 March 1953)

As you can see, the communist ideal was so powerful, so 'sacred' in the words of the otherwise entirely atheist editorial of *France nouvelle*, that it defeated death itself, and justified laying down one's life without fear or remorse. It is no exaggeration to say that here was a new version of the doctrine of salvation. Even today, as a last vestige of this religion without Gods, there are national anthems which extend this hope to their citizens, provided that they sacrifice their destiny as individuals to the higher cause, since 'to die for the homeland is to enter into eternity'.

Of course, we can also find on the right of the political spectrum equivalent forms of patriotism which go by the name of 'nationalism' – the notion that it is worthwhile to lay down one's life for the nation of which one is a member.

In a style that is fairly close to communism and nationalism, scientism furnished its followers with reasons for living and dying. If you have ever read Jules Verne, you will recall the degree to which 'scientists, explorers and builders' (as they used to be called when I was at primary school) are convinced that by discovering an unknown land or a new scientific law, or by inventing a machine for exploring the sky or the sea, they are inscribing their names in the eternity of historical progress and thereby justifying their entire existence. Good for them.

I remarked a few pages earlier that I have always found these new religions absurd – sometimes grossly so. Communism and nationalism caused the deaths of many people, it is true, but it is also their naivety that disconcerts me. The evidence suggests that salvation of the individual life is not the same thing as the salvation of humanity as a whole. Even if we devote ourselves to a 'higher' cause, in the conviction that the ideal is infinitely superior to the individual, it remains true that in the end it is the individual who suffers and dies. Faced with the entirely personal nature of death, communism or nationalism or scientism (or any other -isms we might substitute for them) strike me as desperately empty abstractions.

As the great 'postmodern' thinker, Nietzsche was to ask: is not the passion for 'grand designs' that are supposedly superior to the mere individual, superior even to life itself, merely the final ruse of those religions that we hoped we had left behind? And yet, however derisory these last-ditch attempts at a doctrine of salvation may seem, they represented nonetheless a revolution of considerable scale. For what was hatched by these false religions and their platitudes was nothing less than the secularisation or humanising of the world. In the absence of cosmic or religious first principles, humanity came to be endowed with sacred properties. After all, no one can deny that humanity in its global aspect is, in a sense, superior to the sum of the individuals who compose it, just as the general interest must in principle prevail over that of individuals. This is clearly the reason why these new godless doctrines of salvation succeeded in convincing and converting so many.

But modern philosophy also succeeded, and far more profoundly, in arriving at a different way of formulating the question of salvation.

It was Kant, in the wake of Rousseau, who first launched the notion of 'enlarged thought' to make sense of human life. Enlarged thought was for Kant the opposite of a narrow-minded spirit; it was a way of thinking which managed to disregard the subjective private conditions of the individual life so as to arrive at an understanding of others. To give a simple example, when you learn a foreign language you come to establish some distance both from yourself and from your particular point of origin – that of being English, for example. You enter into a larger and more universal sphere, that of another culture, and, if not a different humanity, at least a different community from that to which you belonged formerly, and which you are now learning not to renounce but to leave behind. By uprooting ourselves from our original situation, we partake of a greater humanity. By learning another language, we can communicate with a greater number of human beings, and we also discover, through language, other ideas and other kinds of humour, other forms of exchange with individuals and with the world. You widen your horizon and push back the natural confines of the spirit that is tethered to its immediate community – this being the definition of the confined spirit, the narrow mind.

Beyond the particular example of languages, the whole realm of human experience is open to you. If to know is to love, then it is also true that by enlarging your horizons and improving yourself, you enter a dimension of human existence which 'justifies' life and gives it a meaning and a direction.

What is the purpose of 'growing up', we are sometimes tempted to ask, and what idea could save us? Let's say that it at least gives a sense to the business of facing death, and we shall return to this 'enlarged' thought later, to flesh it out as it deserves, and to give you a better idea of how it took over from the ancient doctrines of personal salvation. But for the moment we must pass through another stage: that of 'deconstruction' – the critique of pre-existing constructions of the world, their illusions and naiveties. And at this point

Nietzsche enters – the master of suspicion, the most abrasive thinker – a man who marked a turning point, philosophically speaking, for all that came afterwards. It is time for us to understand why this was the case.

5
POSTMODERNITY: THE CASE OF NIETZSCHE

In contemporary philosophy, we call 'postmodern' those ideas which, from the mid-nineteenth century, were to set about dismantling the humanist creed of modernity, in particular the philosophy of the Enlightenment. In the same way that the latter broke with the grand cosmologies of Antiquity and brought about a new critique of religion, so too postmodernity was to set about demolishing the two strongest convictions of the Moderns from the seventeenth to the nineteenth centuries: the belief that the human individual is at the centre of the world – which came to form the basis of all moral and political values; and the belief that reason is an irresistible force for emancipation and that, thanks to the progress of 'Enlightenment', we are going to become ever freer and happier.

Postmodern philosophy contested both of these ideas. It was to offer both a critique of humanism and a critique of rationalism. And, without any doubt, it is with Nietzsche that postmodernity arrived at its zenith. While there remain many reservations concerning Nietzsche, the radical aspect and the violence of his assault upon the idols of modernity are equalled only by the genius with which he was able to marshall his forces.

But – as the great contemporary philosopher Heidegger

asked – why this need to pull down or 'deconstruct' what modern humanism had taken so much trouble to erect? Why turn yet again from one vision of the world to another? On what grounds did the gains of the Enlightenment come to seem insufficient or illusory; what reasons could seem important enough to provoke modern philosophy to want to 'go even further'?

The answer is quite simple, if we stick to essentials. Modern philosophy, as we have seen, had deposed the *cosmos* and turned its back on religious authority, replacing them with reason and individual freedom; the democratic and humanist ideal of moral value founded upon man's humanity to man – based on what made man different as a species from all other animals. However, as we have also seen, this was achieved on the basis of radical doubt introduced by Descartes. In other words, by making the critical spirit 'sacred', a freedom of thought which went so far as to make a *tabula rasa* of the entire past, its intellectual legacy and traditions. Science was itself so thoroughly imbued with this spirit that nothing could stop it in its quest for truth. Like the sorcerer's apprentice who unleashes forces which soon escape his control, Descartes and the Enlightenment philosophers unleashed a critical spirit which, once in motion could not be stopped, somewhat like an acid that continues to eat into the materials with which it comes in contact, even after water has been thrown over it.

As we have seen, modern science, the fruit of the critical spirit and of scientific method, laid waste to the preceding cosmologies and greatly weakened, initially at least, the foundations of religious authority. This is a fact. Even so, as we saw at the end of the preceding chapter, humanism was far from dismantling the underlying religious assumptions: the opposition between the here and the hereafter, of paradise as opposed to earthly reality, or, if you prefer, of the ideal set against the real. This is why, in Nietzsche's eyes, even if the republicans who inherited the mantle of Enlightenment pronounced themselves atheists, or materialists, they continued in effect to be *believers*! Not, of course, in the sense that they still prayed to God, but in the sense that they revered their new illusions, since they continued to believe that certain values were superior to life itself and that we must transform reality to make it conform to higher ideas: the rights of man, science, reason, democracy, socialism, equal opportunity and so on.

Now this vision remains fundamentally theological, even if it does not realise the fact and thinks of itself as revolutionary or anti-religious. Briefly, to postmodern eyes, and for Nietzsche above all, Enlightenment humanism remained a prisoner of the underlying religious structures. Which is why modernity was going to have to endure the same critique that it had unleashed upon the supporters of cosmologies and religious belief systems.

In the preface to *Ecce Homo*, one of his rare works which takes an overtly confessional form, Nietzsche describes his philosophical attitude in terms which describe perfectly his rupture with modern humanism. The latter was still proclaiming its belief in progress, its conviction that the diffusion of science and technology would bring happier days and that history and politics must be shaped by an ideal. This was precisely the type of belief, this godless religion, or – as Nietzsche expresses it in his very idiosyncratic vocabulary – the type of 'idol' which he proposed to deconstruct, by 'philosophising with a hammer'. Let us listen to him for a moment:

> Improve mankind? That is the last thing that *I* of all people will promise to do. Don't expect new idols from me; let the old idols learn what it costs to have feet of clay. To *overthrow idols* – my word for 'ideals' – that rather is my business. Reality has lost its value, its meaning, its veracity, and an ideal world has been *fabricated* to take its place . . . The *lie* of the ideal has hitherto been the curse on reality, through which mankind itself has become mendacious and false down to its deepest instincts – to the point of worshiping the *opposite* values to those which alone could guarantee it prosperity, a future, the exalted *right* to a future. (*Ecce Homo*)

It is no longer a case therefore of piecing together or reconstituting a human world, Kant's 'realm of ends', where men are at last equal in dignity. To postmodern eyes, democracy, whatever content it is assigned, is merely one more religious illusion among others; one of the worst, in fact, since it usually dissimulates itself under the appearance of a break with the religious world, pretending to a free 'laicity' (the state of being of the people, not members of the clergy). Nietzsche never ceases to return to this, with the greatest lucidity, as in this passage from *Beyond Good and Evil*:

> We who hold to a different belief – we who consider the democratic phenomenon to be not merely a decadent

> form of political organisation, but a decadent (that is to say, diminished) form of the human being, one that reduces him to mediocrity and debases his value – where are we to pin our hopes?

Not on democracy, clearly! It is undeniable: Nietzsche is wholly against democracy, and, unfortunately, it is not entirely by chance that he was taken up by the Nazis as one of their inspirations. But if we want to understand Nietzsche, before condemning him, we must go further: if he abhorred ideals as such, if he wanted to smash the idols of modernity with his philosophical hammer, it is because they all derive from a negation of life, from what he termed 'nihilism'.

Before advancing further, we need to understand this central tenet in Nietzsche's deconstruction of modern moral and political utopias. It was Nietzsche's conviction that all ideals, whether explicitly religious or not, whether coming from the right or the left, conservative or progressive, spiritualist or materialist, possessed the same configuration and the same purpose: fundamentally, they are all the product of a theological world-view, because they all persisted in assuming a hereafter that is better than the here and now, in offering values supposedly superior and external to life itself; or, in philosophical terms, values that are 'transcendental'. In Nietzsche's eyes, such a fabrication was always animated, covertly of course, by 'wicked intentions'; its true purpose being not to help humanity, but only to judge and finally condemn life itself; to deny actual truth in the name of false realities, instead of accepting the real as it is.

This negation of the real in the name of the ideal

was what Nietzsche meant by 'nihilism'. Thanks to this fiction of supposed ideals and utopias, we place ourselves beyond reality, beyond life, whereas the heart of Nietzsche's thought is that there is no transcendence, that all judgement on life is a symptom, a product of life, and can never situate itself outside of life. If you can grasp this, nothing can hold you back from reading him: that there is nothing outside this reality, no beyond, no above, neither in heaven or in hell; and all the fine ideals of politics, ethics and religion are merely 'idols', metaphysical projections, fables that turn their back on life prior to turning against life.

You can see why postmodern philosophy was destined inevitably to criticise the Moderns as being excessively in thrall to religious utopia. The Moderns invented the critical spirit, the practice of doubt, the lucidities of reason . . . only for all the weapons essential to their armoury to be turned against them. The principal postmodern thinkers, Nietzsche of course, but also to varying degrees Marx and Freud, have been justly described as 'masters of suspicion': their purpose is to deconstruct the illusions with which classical humanism deluded itself. These philosophers adopted as a first principle the sixth sense that, behind the curtain of traditional beliefs and 'good old-fashioned values' which pretend to beauty, truth and transcendence, lurks always concealed interests, unconscious choices, and deeper (for the most part inadmissible) truths. As with psychoanalysis, postmodern philosophy learnt above all else to distrust self-evidence, received ideas; to look behind, above and sideways if necessary to bring to light the hidden agendas which underpin all values.

This is why Nietzsche dislikes grand solutions, and 'consensus', and why he prefers shortcuts, sidelines, contention. Like the founding figures of contemporary art, such as Picasso or Schoenberg, Nietzsche is an avant-gardist, someone who intends above all to innovate and make a clean sweep of the past. What was to define the postmodern mood, above all else, was its irreverence, its impatience with fine sentiments and bourgeois values: everyone who prostrated themselves before scientific truth, reason, Kantian morality, democracy, socialism, republicanism. The avant-gardists, with Nietzsche in the lead, took it upon themselves to smash everything, so as to expose to full view what was concealed behind. They were, you might say, hooligans (albeit sophisticated ones)! As far as they were concerned, humanism had lost all its creative and destructive energies: this attitude explains the radicalism, the brutality and even the more frightening aspects of postmodern philosophy. Yes, we must acknowledge without contention that it is no accident that Nietzsche became the cult philosopher of the Nazis, in the way that Marx became so for Stalinists and Maoists. However, Nietzsche's thought, intolerable at times, is also inspired. One might not share his ideas – one might even detest some of them – but one cannot think in the same way after reading his work. This is a sure sign of genius.

To indicate the principal features of his philosophy, I shall continue to follow the three grand axes of enquiry to which we have become accustomed: *theoria*, *praxis*, and doctrine of salvation.

Some admirers of Nietzsche believe that it is pointless to try and find anything as organised as a *theoria* in the writings of one who was beyond all others the destroyer of rationalism and a tireless critic of what he called 'theoretical man' – driven by 'the passion for knowledge'. It seems sacrilegious – it would have made Nietzsche laugh – to search for a 'morality', given that Nietzsche never stopped describing himself as an 'immoralist', or to seek wisdom in the works of one who died insane. And what sort of doctrine of salvation might we expect from a thinker who had the audacity to compare himself to the Antichrist and to deride all forms of 'spirituality'? In reply to which I would say, do not listen to everything you are told, and always judge for yourself. Read the works of Nietzsche – starting with *Twilight of the Idols*, particularly the brief chapter entitled 'The Problem of Socrates'. Then make up your own mind.

It is as plain as day, from the first reading, that you will not find in Nietzsche a *theoria*, a *praxis*, or a doctrine of salvation in the sense that we have encountered in discussing the Stoics, the Christians or even Descartes, Rousseau and Kant. Nietzsche is truly what might be termed a 'genealogist' – it is the name he gave himself – who spent his life dismantling the illusions of traditional philosophy.

Does this mean, then, that you will not find in his work a body of thought which takes the place of the ancient certainties it comes to bury, and which substitutes for the 'idols' of traditional metaphysics? As we shall see, Nietzsche does not deconstruct Greek cosmology, Christianity or Enlightenment philosophy merely for the pleasure of destruction; he is clearing the way for radical new thoughts which truly constitute a new *theoria*, praxis and even philosophy of salvation. It is nothing less than a new way of thinking. In which case it remains a philosophy.

A 'Gay Science': Free from Cosmos, God and the 'Idols' of Reason

Let us remind ourselves of the two key aspects of philosophical *theoria*. There is the *theion* and the *oraio*; the *divine* that we are seeking to locate within the real, and the act of *seeing* that contemplates it: there is *that which* one tries to understand and *that with which* one tries to accede to understanding (the instruments one employs to get there). Theory always combines a definition of the essence of being, of what is most important in the world around us (what we call *ontology* – *ontos* deriving from the Greek word for *being*) and a definition of *vision*, the means of apprehension which will enable us to grasp it (what we refer to as a theory of knowledge).

We shall trace these two components of *theoria* in the thought of Nietzsche, in order to see the distortions to which he subjects them, and how he overhauls them in an unprecedented fashion. As you shall see, his *theoria* is in fact an '*a-theoria*' – in the sense that one says of a man who does not believe in God that he is *a-theist*: literally, without God (the Greek prefix '*a*-' meaning 'without'). Because for Nietzsche, the innermost essence of being partakes neither of cosmos nor divinity; knowledge itself is no longer a category of *vision*, analogous to the Greek *orao*. It is not an act of contemplation, or a passive spectacle, as it was for the

Ancients. Nor is it, as it was for the Moderns, an attempt to elaborate against all odds the sum of relations between phenomena, so as to discover a new order and a new meaning. To Nietzsche, knowledge is an act of 'deconstruction', hence the name 'genealogy', as mentioned earlier. The word is eloquent: as with the activity that consists of tracing the different branches of a family tree, true philosophy according to Nietzsche brings to light the hidden origins of values and ideas which pretend to be untouchable, sacred and handed down from on high, so as to bring them down to Earth and disclose their nature (and all too often *earthly* – one of Nietzsche's favourite words – origins).

A Theory of Knowledge: Genealogy Replaces Theoria

Nietzsche's most profound insight, and one that will underpin his entire philosophy is that there does not exist, categorically, any perspective external to or higher than life itself, any point of view privileged enough (for whatever reason) to abstract itself from the tissue of forces which are the ground of the real, and are therefore the innermost essence of being. Consequently, no judgement on existence in general has any sense, other than illusory, or symptomatic of the condition of the vital forces of the individual concerned.

This is how Nietzsche sets out his argument in a decisive passage from *The Twilight of the Idols*:

> Judgements, value-judgements on life, whether for or against, can ultimately never be true: they have value only as symptoms, they can be considered only as symptoms – in themselves such judgements are foolish. We must really stretch out our fingers and make the effort to grasp this astonishing refinement: *that the value of life cannot be assessed*. Not by a living person because he is an interested party, is indeed even the object of dispute, and not the judge; nor by a dead person, for a different reason. For a philosopher to find the *value* of life problematic is therefore an objection against him, a question mark against his wisdom, a piece of unwisdom. (II, 2)

For the deconstructionist, for the genealogist, there can be no 'objective' or 'disinterested' value judgements – independent of the vital interests of the speaker – which devastates the classical conceptions of law and ethics – and there can be neither autonomous and disinterested judgements, nor objective and universally valid 'facts'. All our judgements, all our utterances, all the sentences we employ, all our ideas, are expressions of our vital energies, emanations of our inner life and in no sense abstract entities, autonomous and independent of the forces within. The whole project of genealogy is to prove this new truth.

According to one of Nietzsche's most celebrated statements, 'There are no facts, only interpretations'. In the same way that we can never be autonomous and free individuals, transcending the real at the heart of which we must live our lives, but are solely the product of historical forces, entirely immersed in the reality that is ours – by the same token, and contrary to what is claimed by the positivists or scientists, there are no 'facts in themselves'. The scientist always says: 'These are the facts!' – whether to remove an objection or merely to express what he feels, faced with the constraints of 'objective truth'. But the 'facts' to which he claims to submit, as if to an abstract and incontrovertible reality, are merely – on a deeper level – the product (itself changeable) of history, and of the forces that comprise life at a particular moment.

True philosophy leads us towards an abyss: the deconstructive activity of the genealogist ends in the realisation that underlying the business of judgement there is no foundation but a void: behind the 'other-worlds' of traditional philosophy there recede yet more other-worlds, for ever imperceptible. Alone and cut off from the 'herd' of society, the true philosopher must undertake henceforth the agonising task of facing into the abyss:

> Indeed the hermit . . . will doubt whether a philosopher is even *capable* of having 'final and true' opinions, whether at the back of his every cave a deeper cave is lying, is bound to lie – a wider, stranger, richer world behind every surface, an abyss beneath his every depth, and beneath his every abyss an inmost depth. 'Every philosophy is a façade-philosophy' – such is the hermit's judgement . . . Every philosophy also *conceals* a philosophy; every opinion is also a hiding-place; every word is also a mask. (*Beyond Good and Evil*, 289)

If knowledge can never reach absolute truth, if it is for ever pushed back, from one horizon to the next, without ever touching down on solid ground, this is because the real itself is a chaos which no longer resembles the harmonious order of the Ancients or the more or less 'rationalisable' universe of the Moderns. Through this new idea we penetrate to the heart of Nietzsche's thought.

The World as a Chaos Without Cosmos or Divinity

If we want to fully understand Nietzsche, we have to start from the idea that he imagines the world in a manner almost directly opposed to the Stoics. Nietzsche considered the world – organic and inorganic, within us as much as outside of us – to be a vast field of energies, a web of forces and drives whose infinite and chaotic multiplicity cannot be reduced to unity. In other words, the *cosmos* of the Greeks was in his eyes the supreme untruth – a pretty piece of make-believe, with no purpose other than to console and reassure us:

> And do you know what 'the world' is to me? Shall I show it to you in my mirror? This world: a monster of energy without beginning or end, a rigid quantum of forces, unyielding as bronze, becoming neither greater nor smaller, that does not expend itself but only transforms itself . . . a sea of forces flowing and rushing together, in perpetual flux. (*The Will to Power*, 1067)

Of course the *cosmos* of the Greeks had already been exploded by the Moderns, by Kant and Newton, so how can Nietzsche take things further in dismantling the idea of universal harmony? The most succinct answer is that while Kant or Newton strove to find a coherence, an order in the world, by attempting to inject it with rationality, with logic – remember Claude Bernard and his rabbits – for Nietzsche such an enterprise was an utter waste of time and effort. It remained trapped inside fantasies of reason, meaning and logic, because no unification of the chaos of natural forces is possible. Like the Renaissance thinkers who saw the *cosmos* collapsing beneath the blows of the new physics, we are now in the grip of terror, and 'consolation' is no longer possible:

> The world has once again become infinite to us . . . Once again the great shudder is upon us – but who would want to start deifying all over again in the old manner this monster of an unknown world? . . . Alas, too many ungodly possibilities of interpretation are included in this Unknown; too much devilry, stupidity, foolishness of interpretation. (*The Gay Science*, 374)

The scientific rationalism of the Moderns is a mere illusion, and is no more than a way of keeping faith with the illusions of the Ancients – an all too human 'projection' (Nietzsche was already using terms that were soon to be adopted by Freud); in other words a way of substituting our desires for realities, a way of procuring for ourselves the semblance of power over inanimate nature, multiform and chaotic, which in reality escapes our grasp on all sides.

I mentioned Picasso and Schoenberg earlier, as founders of contemporary art who are fundamentally attuned to Nietzsche. If you look at these paintings or listen to this music, you will see that it too delivers us up to a world that is destructured, chaotic, fragmented, alogical, deprived of the 'beautiful unity' which perspective and the rules of harmony conferred upon works of art in the past. This will give you an accurate image of what Nietzsche was attempting to think towards, fifty years earlier – and it is worth noting that philosophy, even more so than the arts, is again ahead of its time.

In these circumstances there was little chance for philosophy to stay in the business of contemplating a divinely ordained universe, of any variety. It becomes impossible for philosophy to take the form of a *theoria*, in the strictest sense, as a 'vision' of the 'divine'. However, Nietzsche does remain a philosopher. He strives to understand this reality that surrounds us, to grasp the underlying nature of a world within which, even if it is a chaos, we must absolutely learn to situate ourselves.

But rather than trying at all costs to discover a logic to this chaos, this tissue of contradictory forces that is the universe – which he designates as 'Life' – Nietzsche was to distinguish between two quite distinct types of force – or, as he was to say, two 'drives' or 'instincts'; on the one hand, 'reactive', and, on the other, 'active'. It is upon this distinction that his thinking is founded. Reactive forces, on an intellectual level, are modelled upon the same 'will to truth' that animates classical philosophy and science; in politics they attempt to realise the democratic ideal. Active forces, on the contrary, are essentially called into play by art, and their natural sphere is that of aristocracy.

The Negation of the Visible World

Reactive forces: those forces which can only deploy themselves in the world and achieve their full effect by repressing, annihilating or distorting other forces. In simpler terms, they succeed only by opposing; they belong to the realm of 'no' rather than 'yes', of 'against' rather than 'for'. The model here is the classical search for truth, since this always triumphs more or less *negatively*, by setting itself to refute errors, illusions, false opinions. This applies as much in philosophy as in the positivist sciences.

The example Nietzsche uses, and which he has in mind when he speaks of reactive forces, is that of the great dialogues of Plato. You need only be aware here that these dialogues almost always take the following form: the readers – or rather, listeners, for they frequently take place before a public, like theatrical performances – witness an exchange between a central figure, usually Socrates, and his interlocutors, who are sometimes well-disposed and somewhat pliable, sometimes hostile and in a quarrelsome frame of mind; notably so when Socrates takes on those who were called the 'Sophists' – the masters of public speaking, of 'rhetoric'. The Sophists made no attempt, unlike Socrates, to seek the truth, but only to instill the best means of seducing and persuading by the art of oratory.

Having settled on a philosophical theme for discussion, of the 'What is courage?' or 'What is beauty?' variety, Socrates would propose that they survey together the 'commonplaces', the current opinion on the topic in question, as a point of departure, with a view to elevating the discussion, point by point, and if possible arriving at the truth of the matter. Once this opening survey was concluded, discussion could take place; what is known as the 'dialectic', the art of dialogue, in the course of which Socrates proceeds to ask his interlocutors a stream of questions, usually to show that they are contradicting themselves, that their initial ideas or convictions do not hold water, and that they must reflect more if they are to get any further.

You must also know one more thing about Plato's dialogues before we return to the 'reactive forces' of Nietzsche: the exchanges between Socrates and his interlocutors are always in reality unequal. For Socrates always takes up a position *at an angle* to whomever he interrogates. He makes a show of not knowing. He likes to play the innocent – let's say there is an Inspector Columbo side to his personality. But in truth he knows exactly where the interview is heading. This is not a level playing field: Socrates is pretending to be equal, whereas he has the advantage, that of the master over his pupil. It was this that the German Romantics christened 'Socratic irony' – because Socrates is playing a game; he is not merely at an angle to those who surround him, but above all towards his own beliefs, since he is perfectly aware – contrary to appearances – that he is playing a role.

And it is this attitude that Nietzsche considered to be essentially *negative or reactive*: not only does the truth which Socrates is seeking reveal itself only in terms of a refutation of others, but Socrates himself affirms nothing: he takes no risk, reveals nothing, proposes nothing positive. He merely contents himself with placing his interlocutor in difficulty, leading him to contradict himself, as a way of inducing him to give birth to the truth.

In one of his dialogues, Socrates describes himself as

a torpedo fish – he paralyses his prey. For it is in *refuting* others that the dialogue advances, in order to arrive at a better idea of things. You will now see the link which exists in Nietzsche's mind between the Socratic passion for the true, the will to find the truth – whether philosophical or scientific – and the idea of 'reactive' forces. For Nietzsche, the search for the truth reveals itself to be *doubly* reactive: true knowledge is not to be had solely through a combat against error, bad faith and untruth, but more generally, through a combat against the illusions inherent in the sensible world. Philosophy and science are only able to function in effect by opposing 'the intelligible world' to the 'physical world' in such a manner that the second is always devalued in relation to the first. This is a crucial point for Nietzsche, and it is important that we understand it fully.

Nietzsche accuses all the grand scientific, metaphysical and religious systems – Christianity in particular – of having systematically 'despised' the body and the senses in the interests of reason and rationality. It might seem strange to you that he puts religion and science in the same basket. But there is no inconsistency here. Despite all that separates and even opposes metaphysics, religion and science, they share in common a claim to accede to ideal truths, to intellectual realities, entities that are not available to the senses, and to notions which do not partake of the corporeal world. It is therefore 'against' corporeal reality – here again the idea of 'reaction' – that these systems strive, because (as everyone knows) we are ceaselessly deluded by them.

To take one example, if we confine ourselves strictly to the evidence of the senses – to sight, touch and so on – water, for example, appears to us in a variety of guises, often contradictory (boiling water, cold rain, soft snow, hard ice etc.), whereas it is always 'in truth' one and the same substance. Which is why we are told we must attempt to rise above the sensible world, and even to think *against* the senses – reactively, as far as Nietzsche is concerned – if we wish to attain 'the intelligible', to apprehend 'the idea of water'.

From the point of view of the 'will to truth', as Nietzsche puts it, the scientist or philosopher who wishes to attain true knowledge must consequently reject all those impulses which rely too exclusively on the evidence of the senses, of the body. In effect, then, philosophy and science would have us mistrust everything that is essential to the creation of art. And Nietzsche's suspicion, of course, is that behind this 'reaction' lurks an agenda whose concern is quite other than the search for truth alone; namely, a hidden prejudice in favour of 'the beyond' as against 'the here and now'.

If we challenge not merely the 'search' for truth, but also the fellow-travelling ideals of democratic humanism, then the critique of modern philosophy and the 'bourgeois values' on which it is based, as far as Nietzsche is concerned, is now complete. For the truths which science endeavours to attain are 'intrinsically' democratic: they are those which claim that one value applies to all, at all times and in all places. A formula such as $2 + 2 = 4$ knows no barriers of social class, of space or time, no frontiers of geography or history; it lays claim to universality. Thus the truths of science are at the heart of humanism; or, as he would prefer to say, they are 'plebeian' and profoundly 'anti-aristocratic'.

Here too, moreover, is what scientists (usually republican, as far as Nietzsche is concerned) prize about science: it addresses the weak as much as the powerful, the poor and the rich, commoners and kings. Nietzsche amuses himself by reminding us of the plebeian origins of Socrates, inventor of philosophy and science, and the first to promote reactive forces idealising 'the truth'. I shall quote a passage from the chapter on Socrates (in *Twilight of the Idols*) which links the will to truth with the legendary ugliness of the hero of Plato's dialogues, who signalled the end of an aristocratic order imbued with 'distinction' and 'authority':

> Socrates belonged by origin to the lowest rung of the people: Socrates was rabble. We know, we can even still see, how ugly he was . . . Was Socrates actually even a Greek? Ugliness is often enough the expression of a stunted development, hampered by cross-breeding . . . With Socrates, Greek taste switched over to dialectics: what is actually going on here? Above all it means that a *noble taste* has been defeated. With dialectics, the rabble comes out on top. Before Socrates, dialectical manners were disapproved of in polite society . . . Whatever needs first to have itself proved to be believed is of little value. Wherever it is still good manners to have authority, and people do not 'reason' but command, the dialectician is a kind of buffoon: he is laughed at, he is not taken seriously. Socrates was the buffoon who *got himself taken seriously*.

It is difficult, today, to ignore what is disagreeable about this passage. All the ingredients of fascist ideology seem to come together: the cult of beauty, a 'distinction' from which the 'mob' are by nature excluded; the classification of the individual according to social origins; the equivalence between the populace and ugliness; the valuing of the nation-state (ancient Greece, in this case); and an unsavoury suspicion about cross-breeding relating to social decadence. Nothing is missing. But let us not judge by first impressions. This account fails to do justice to what is nonetheless profound in Nietzsche's interpretation of the character of Socrates. Rather than reject it outright, we might examine the meaning of his propositions and tease out a deeper significance.

Before doing so, we need to increase our understanding of another aspect of Nietzsche's thought, namely those 'active' forces which balance the reactive, and which together complete his version of the world, his attempt upon reality.

An 'Aristocratic' Vision of the World

Contrary to all that is reactive, the active forces take effect in the world and do their work without needing to disfigure or repress other energies. It is in art, and not in philosophy and science, that these forces find their natural home. Closer examination will allow us to understand Nietzsche's fearsome verdict upon Socrates and see how his 'ontology' fits together, by which I mean his complete account of the world as an ensemble of reactive and active forces.

Contrary to the 'theoretical man' – the philosopher or scientist of whom we have been speaking – the artist is the figure who, above all others, imposes values without discussion, opens up perspectives and invents worlds without needing to demonstrate the legitimacy of his propositions, still less to prove them by a *refutation* of those works which preceded his own. Like the aristocracy, the artist commands without arguing with anyone or anything – and note that it is in this sense that Nietzsche declares: 'Whatever needs first to have itself proved to be believed is of little value.'

Clearly you can like Chopin, Bach, rock music or techno, the Dutch painters or contemporary art, without it occurring to anyone to require that you choose one of these to the exclusion of the others. In the realm of seeking truth, on the other hand, at some point or other you must make a choice: Copernicus is right and Ptolemy was wrong; Newtonian physics is demonstrably truer than that of Descartes. In this way truth establishes itself only by progressively removing the errors which in effect constitute the history of science (with whose corpses the stage is littered). The history of art, on the contrary, is a space where different and even radically opposing works can coexist. Not that tensions and quarrels are absent; aesthetic conflicts have often been of the most violent and passionate nature. Nonetheless, they are not settled in terms of 'who is right and who is wrong' – they remain unresolved, and they always leave open – at least in retrospect – the possibility of an equal result for their respective protagonists. Nobody would dream of saying, for example, that Chopin is right and Bach wrong, or Ravel mistaken in comparison to Mozart. All of which connects with the fact that, since the dawn of philosophy in Greece, two kinds of discourse, two conceptions of words and their use have clashed with each other.

On the one side sits the Socratic and reactive model, which seeks the truth through debate and dialogue, and in order to get there, takes its stand against the various faces of ignorance, stupidity or bad faith. On the other side sits the model of the Sophists, which makes no attempt to seek the truth, but seeks merely to seduce, to persuade, to effect an audience with almost physical intensity, and win over by the power of words alone. The first procedure is that of philosophy and science: where language is solely an instrument in the service of a higher reality, the intelligible and democratic Truth which will one day impose its reign upon each and every one. The second is that of art, of poetry: words are no longer simply means, but ends in themselves, which possess intrinsic value from the moment they produce an aesthetic effect upon those capable of experiencing it.

One of the tactics employed by Socrates, in his oratorical sparring with the Sophists, illustrates perfectly this opposition. Whenever a great Sophist, Gorgias or Protagoras, for example, had just finished a dazzling speech before an audience still under the spell of enchantment, Socrates would feign incomprehension, or, even better, deliberately arrive too late, after the spectacle was over. This provided him with an excellent pretext to ask the speaker to give a summary of his speech – to reformulate, briefly if possible, the salient points. As you can imagine, for the Sophist, this is virtually impossible – which is why Socrates' request is, for Nietzsche, the product of pure malice! As easy to reduce a conversation between lovers to its 'kernel of sense', or ask Baudelaire or Rimbaud to summarise one of their poems! 'The Albatross'? About a bird struggling to achieve lift-off. 'The Drunken Boat'? Concerns a sea-going vessel in difficulties. Socrates has no difficulty in keeping the score: as soon as his adversary commits the blunder of taking up the challenge he is lost, because as far as art is concerned, all the signs suggests that it is not the truth-content that matters but the emotional logic, and the latter, of course, has no defence against the reductivism of a summary.

We begin to see finally what Nietzsche means, in the passage describing 'the ugliness' of Socrates, when he links him with democratic ideology, or when he denounces, a little further on, 'the rabble's resentment' which takes pleasure in wielding triumphantly against its interlocutors 'the knife-thrusts of the syllogism'. Rather than fascist noise, what speaks here is Nietzsche's aversion towards the will to truth (at least in its rationalist and reactive forms – for we must not forget that Nietzsche is himself also a searcher after truth).

Similarly, when he speaks of 'stunted development' and associates the idea of cross-breeding with degeneracy, let us ignore the apparent stench of racism. However ambivalent or distasteful it may appear, he is setting his sights on something profound, referring to a phenomenon which we shall need to clarify: namely, the fact that forces which collide, which constantly thwart each other – what Nietzsche refers to here as 'cross-breeding' – dilute life and make it less intense, less interesting.

In Nietzsche's eyes – perhaps to Nietzsche's ears, rather, since the whole vocabulary of sight, of vision, of *theoria* is so contaminated – the world is not a *cosmos*:

neither a natural order, as it was for the Ancients, nor an order constructed by the will of man, as it was for the Moderns. The world is a chaos, an irreducible plurality of forces, instincts and drives which ceaselessly clash. This being so, the problem presents itself as follows: by their constant jostling, these forces (within us, as much as in the world outside) are in constant danger of thwarting each other, and at worst of creating a blockage, a diminution or weakening. In this state of conflict, therefore, life languishes, becomes less vibrant, less unbounded, less spirited, less strong. In this respect Nietzsche heralds the coming of psychoanalysis, in which our unconscious psychic conflicts and internal battles prevent us from living a full life, make us ill and prevent us from 'playing and working', to use one of Freud's terms.

Many recent commentators on Nietzsche have committed the same grave error in respect of his thinking: they have hastily concluded that, to render life more free and more spirited, Nietzsche proposed to reject the reactive forces in order to give free play to the active forces alone – to liberate the physical and corporeal, by rejecting 'the cold and dry reign of reason'. At first sight this might seem logical enough, but let us remember that such a 'solution' is typical of what Nietzsche called 'stupidity': because, clearly, to reject all reactive forces is merely to founder upon another kind of reaction, since it would in turn mean setting oneself against reality in one of its forms. Nietzsche is not inviting us to follow him into some version of anarchy, or emancipation of the senses, or 'sexual liberation', but is leading us on the contrary towards a more intense, more dialectical experience and mastery of the multiple forces that govern life. It is this that Nietzsche refers to in the phrase 'the grand style'. And it is with this notion that we reach the ethical core of this self-styled immoralist.

Beyond Good and Evil

There is of course something paradoxical about searching for a morality in the thought of Nietzsche – just as there was about searching for a *theoria*. You will recall the violence with which he rejects all attempts to improve the world. He is for ever characterised as the 'immoralist' par excellence, who railed against charity, compassion and altruism in all their guises, whether Christian or otherwise.

As I have said before, Nietzsche detested the notion of the ideal, and was among those who contested the first tentative steps of modern humanitarianism, in which he saw merely a watered-down, feeble version of Christianity:

> To proclaim a universal love of humanity is, in practice, to acknowledge the *preferment* of all that is suffering, ill-constituted, degenerate . . . For the wellbeing of the species, it is necessary for the ill-constituted, the feeble, the degenerate to perish. (*The Will to Power*)

Sometimes, his anti-charitable passion, or his relish for catastrophe, border on delirium. At one point, according to friends, he could not contain his joy when a minor earthquake destroyed some houses in Nice – where he nonetheless liked to spend time – and his dismay that the disaster was not as serious as originally thought. Happily, he learnt shortly afterwards that a major cataclysm had ravaged the island of Java. (Writing to his friend Paul Lanzky: 'Two hundred thousand wiped out at a stroke – how magnificent! What we need is the total destruction of Nice and all who live there').

Is it not therefore an aberration to speak of a Nietzschean 'morality'? And what might it consist of? If human life is merely a web of blind forces tearing each other apart, if our value judgements are merely arbitrary, more or less compromised according to the case, but necessarily devoid of any significance other than as symptoms of our vital spirits, why would we expect any ethical consideration whatsoever from Nietzsche?

One hypothesis, it is true, which has appealed to certain 'leftist' Nietzscheans (an unlikely category, perhaps, and one that would have driven Nietzsche even closer to the edge) runs as follows: if the reactive forces may be thought of as 'repressive', and the active forces as progressive and emancipatory, must we not simply overcome the former in the interest of the latter? Should we not go further and proscribe all norms, insist that it is 'forbidden to forbid' (to use a 1968 student motto), that bourgeois morality is the invention of clerics – and so forth, so as to liberate those drives which are operative in art, in the body, in our emotions?

So it might seem. Certainly, Nietzsche was read in this lurid light during the student revolutions of 1968: as a rebel, an anarchist, an apostle of sexual liberation and the emancipation of the body. But it suffices merely to read Nietzsche to see that this hypothesis is not merely simplistic and absurd, but directly antithetical to his beliefs. That Nietzsche was anything but an anarchist, and that he insisted upon this fact loudly and clearly, can be seen for example in this passage:

> When the anarchist, as the mouthpiece of social interests *in decline*, waxes indignant and demands 'rights', 'justice', 'equality', then he is merely feeling the pressure of his lack of culture, which is incapable of equipping him to understand *why* he is in fact suffering, and *in which respect* his life is impoverished . . . There is a powerful causal drive within him: someone must be to blame for feeling bad . . . And waxing indignant makes him feel better, too: all poor devils take pleasure in cursing, it gives them a little rush of power. (*Twilight of the Idols*, IX, 34)

One might disagree with Nietzsche's analysis here, but we cannot make him bear the responsibility for the libertarian passion and idealist indignation of a phenomenon such as the Paris riots of May 1968, which he would undoubtedly have considered a prime example of 'the herd instinct'. Whatever our sympathies, we cannot deny his aversion to all forms of revolutionary ideology, whether socialist, communist or anarchist.

That the simple-minded idea of 'sexual liberation' would have frozen him with horror is equally evident: what a true artist, a writer worthy of the name, must seek above all is economy. According to one strand of Nietzsche's thought, 'Chastity is the artist's economy', which he must practise constantly, since 'the force that is expended in artistic creation is the same as that expended in the sexual act: there is only one kind of force' (*The Will to Power*). Besides, Nietzsche does not have words strong enough to proscribe the flood of emotionalism that characterises modern life in the wake of Romanticism, and which he regards as catastrophic.

We must therefore read Nietzsche, before pronouncing on his views or making him the mouthpiece of our own. If we wish to understand him, we must add this rider, which will be clear to any true reader of his work: that any 'ethical' attitude which consists in rejecting some part of our vital energies – corresponding to the reactive forces – in favour of another, wholly 'active' though this may be, is inevitably and by definition counter-productive. This is not merely a corrective to his definition of reactive forces as mutilating and castrating, but is also an explicit and constantly reiterated thesis, as seen in this passage from *Human, All Too Human*:

> Let us suppose a man who loved the plastic arts or music as much as he was moved by the spirit of science [is seduced therefore by both visages of force, active and reactive], and who deemed it impossible to resolve this contradiction by destroying the one and completely unleashing the other power; then, the only thing remaining to him would be to make such a great edifice out of himself that both powers can inhabit it, even if at opposite ends; between which are sheltered conciliatory powers, provided with the dominant strength to settle, if need be, any quarrels that break out. (*Human, All Too Human*, I, 276)

It is this reconciling that is, for Nietzsche, the new ideal, the ultimately credible ideal. Because, unlike all others before, it is not at a false remove from life; it is, on the contrary, explicitly lashed to the cargo of life. And it is this, precisely, that Nietzsche refers to as 'grandeur' – a key term for him – the sign of the 'edifice' of culture, at the heart of which opposing forces, because they are finally harmonised and hierarchised, attain the greatest intensity as well as the most perfect elegance. It is only through this harmonising of opposing forces, even the reactive ones, that our human powers can flourish and life cease to be diminished, mutilated. Thus, wherever a great civilisation developed, whether one thinks of an individual or an era, 'it was its task to force opposing forces into harmony through an overwhelming aggregation of the remaining, less irreconcilable powers, and yet without suppressing or shackling them' (*Human, All too Human* I, 276).

To the question 'What is Nietzschean morality?', then, the following is one answer: the good life is the most intensely lived because it is the most harmonious, the most elegant life (in the sense that one speaks of a mathematical solution which does not take unnecessary detours, or expend needless energy). Which is to say, a life in which the vital forces, instead of acting against each other, tearing each other apart and thereby cancelling each other out, instead learn to cooperate with each other, under the mandate of the active forces rather than of the reactive forces. And this, according to Nietzsche, is 'the grand style'.

On this point Nietzsche's thought is utterly clear, and his definition of 'grandeur', throughout the mature writings, is of an unwavering consistency. As is explained very well in a fragment from his posthumous work, *The Will to Power*, 'the greatness of an artist is not measured by the "fine sentiments" that he excites', but resides in 'the grand style', which is to say his capacity 'to master the chaos within himself, to compel his chaos to become form: to become logical, simple, unambiguous, mathematical, to make oneself the law – that is the grand ambition.'

It needs to be said again, that those who are surprised by these texts commit the error, as inane as it is commonplace, of seeing in Nietzsche a purveyor of anarchism, of 'leftist' slogans which anticipate the libertarian movements of our own time. Nothing is more false, since the virtues of 'mathematical' precision, of clear and rigorous reasoning, have their important part to play in the tangle of life-forces. Let us recall once again the reasoning: if we acknowledge that 'reactive' forces are those which can only operate by denying other kinds of force, we must also agree that the critique of Platonism, and more generally of moral rationalism under all its forms, however justified this might be in Nietzsche's eyes, cannot lead to a pure and simple elimination of rationality. Such an eradication would itself be 'reactive'. We must, if we are to arrive at that grandeur which is the sign of a successful fusion of life's forces, harness them in such a way that they cease to block each other. And, in such a hierarchy, rationality must also find its place.

Nothing can be excluded, and in the conflict between reason and the passions, the latter cannot be privileged to the detriment of the former, without sinking into 'stupidity', as Nietzsche repeatedly insists: 'All passions have a period in which they are merely fateful, in which they draw their victims down by weight of stupidity – and a later, very much later one, in which they marry the spirit, 'spiritualise' themselves. (*Twilight of the Idols*, V, 1)

As surprising as it may seem to libertarian readers of Nietzsche, it is precisely this 'spiritualising' that he converts into an ethical category, and which allows us to accede to a 'grand style' by enabling us to harness the reactive forces instead of 'stupidly' rejecting them – together with all that is to be gained by integrating this 'enemy within', instead of exiling it and as a result weakening ourselves. Nietzsche expresses this in a most straightforward manner:

> The spiritualisation of sensuality is a great triumph over Christianity. A further triumph is our spiritualisation of *enmity*. This consists in our profound understanding of the value of having enemies: in short, our doing and deciding the opposite of what people previously thought and decided . . . Throughout the ages the church has wanted to destroy its enemies: we, the immoralists and anti-Christians, see it as to our advantage that the church exists . . . Even in the field of politics, enmity has become spiritualised. Almost every party sees that self-preservation is best served if the opposite number does not lose its powers. The same is true of *Realpolitik*. A new creation, such as the new Reich, needs enemies more than it does friends: only by being opposed does it feel necessary; only by being opposed does it *become* necessary. Our behaviour towards our 'inner enemy' is no different: here, too, we have spiritualised enmity; here, too, we have grasped its *value*. (*Twilight of the Idols*, V, 3)

In this context, Nietzsche (the self-styled Antichrist and unremitting enemy of Christian values) does not hesitate in asserting, loud and clear, that 'the continuation of the Christian ideal is entirely to be desired' because it offers us, through confrontation, a sure means of becoming greater:

> I have declared war on the anaemic Christian ideal (along with all those things closely associated with it), not with the intention of destroying it, but simply to put an end to its *tyranny*, and to clear the ground for new ideals, *more robust* ideals . . . The continuation of the Christian ideal is one of the most desirable things there is – if only for the sake of those ideals that wish to show their worth alongside it, or even above it – for they need adversaries, and *strong* adversaries, if they are to become strong. Which is why we immoralists need the *power of morality*: our instinct for self-preservation wants our *enemies* to stay strong – it merely wants to achieve *mastery over them*. (*The Will to Power*)

If we have understood the distinction between reactive and active forces, the above passages from Nietzsche – so obscure and contradictory to novice readers – become intelligible. And of course it is 'grandeur' which constitutes the beginning and end of 'Nietzschean ethics', and which should guide us in our search for the good life, and for a reason which becomes gradually clearer: because it alone enables us to integrate all forces within ourselves, thereby authorising us to lead a life that is more *intense*, more richly diverse, but also more 'powerful' – in Nietzsche's sense of 'the will to power' – because it is more harmonious. Harmony here is not the harmony of the Ancients, that condition of peaceful contentment, but harmony as the most vigorously tested strength, deriving from the avoidance of those conflicts which exhaust us and the self-hurt which depletes us.

The Will to Power

The notion of a 'will to power' is so central that Nietzsche places it at the heart of his definition of the real, the crowning point of what we have called his 'ontology'. Or, as he repeatedly describes it, the will to power is 'the innermost essence of Being'.

Here we must avoid a major and frequent misunderstanding: the will to power has nothing to do with a lust for power in the world, a desire to occupy some important position or other. It refers to something quite different. It is the will to intensity of experience, the will to avoid at all cost the internal wrenchings that I have described, which by definition diminish us, so that our powers cancel each other and the life inside us stagnates and weakens. The will to power is not the will to conquer, to have money and influence, but a profound desire for a maximum intensity of life, for a life that is no longer impoverished and torn apart by self-division, but on the contrary lived to the full.

By way of example, let us consider the feeling of guilt, when, as the phrase goes: we 'hold something against ourselves'. Nothing is worse than this internal conflict; this condition from which we are unable to find an exit, which paralyses us to the point of removing all joy. And we must think too of the thousands of minor 'unconscious feelings' of guilt, which pass unnoticed, but which produce their own equally devastating effects upon our 'powers'.

Rather as, in certain sports, we can be said to 'pull our punches' rather than 'let fly' – in deference to some buried remorse, an unconscious fear inscribed in the body.

The will to power is not the will to have power, but, as Nietzsche also phrases it, 'the will to will' – the will that seeks to exercise itself, and which is not enfeebled by internal strife, guilt and unresolved conflicts, but which realises itself in 'grand style', in a version of life in which we have done finally with fear, remorse and regret – all internal conflicts which 'weigh us down' and prevent us living with a lightness of being. Let us examine in detail what this might mean.

A Concrete Example of the 'Grand Style'

We have only to think of what must be done when practicing a difficult sport or art to arrive at a perfect execution. Think of the arc of the bow along the strings of a violin, the fingers on the chords of a guitar, or a serve in tennis. When you observe the trajectory of a shot played by a champion, it is of a disconcerting simplicity and facility. Without apparent effort and a graceful fluidity, the player dispatches the ball with astounding velocity: the forces in play in performing this gesture are perfectly integrated. They are in perfect harmony, are fully coordinated, without division, without loss of energy, without consequent 'reaction' in Nietzsche's sense of the term. The consequence of which is an admirable reconciliation of grace and power as can be seen already in the very young, provided they are endowed with a little talent.

On the contrary, the player who has started too late will display an incurably chaotic movement, uncoordinated, or, as we say 'hampered'. He pulls his punches rather than following through . . . And he never stops criticising himself, muttering self-accusations each time he misses a shot. Conflicted at every turn, he plays against himself rather than his adversary. Not only has his rhythm departed, but his power has disappeared: for the simple reason that the forces in play, instead of co-operating, thwart and block each other.

This is what Nietzsche proposes to go beyond, in the moral life. Which is not to say that he is proposing a new 'ideal', a new idol – which would be self-contradictory, for the model he sketches is, unlike all previously constructed ideals, wedded to life itself as it unfolds. It does not in any sense aspire to 'transcendence', located above or beyond the present, in some superior or exterior relation to it. Rather it is about imagining to oneself what a life might be that took as its model 'the free gesture', the gesture of a champion or an artist which unites the greatest diversity to achieve the greatest compression or harmony of forces, without laborious effort, without loss of energy. This is in effect the 'moral vision' of Nietzsche, in whose name he denounces all 'reactive' versions of morality which, since Socrates, have extolled a resistance to life, a lessening of life.

Opposed to 'the grand style', therefore, are ranged all those habits which work against mastery of self – a mastery made possible only by harmonising and hierarchising the chaos of forces within us. In this respect, the unleashing of passions which certain 'liberationist' creeds have tried to promote represents the worst of worlds, since it always involves an internal and reciprocal conflict of forces and a consequent ascendancy of all that is reactive.

Such a mutilation of self exactly defines what Nietzsche refers to as 'ugliness'; the latter manifesting itself whenever passions that are unleashed jostle and weaken each other. 'When there is contradiction, and insufficient coordination of internal desires, there is a diminution of the organising power, of the will . . .' (*The Will to Power*); under which conditions, the will to power languishes and joy gives way to guilt and resentment.

The example I gave of 'the grand style', as the reconciling of active and reactive forces which alone permits access to the 'power' within us – the backhand return of a tennis player – is not Nietzsche's example. But he has many other images which illustrate the idea, and you should be familiar with at least one of these, which he regarded as the most important. It concerns the opposition between classicism and romanticism.

To simplify matters, we could say that classicism refers to ancient Greek art, but equally to French art of the seventeenth century, whether the drama of Molière or Corneille, or the art of landscape gardening, with its trees shaped into geometrical patterns. When you visit the wing of a gallery or museum dedicated to antique sculpture, you will notice that the Greek statuary – perfect embodiments of classical art – has two dominant and typical characteristics: the figures are consummately proportioned, as harmonious as one could wish, and the faces are absolutely calm and serene. Classicism is a style which accords pride of place to harmony and reason. It deeply distrusts the unleashing of emotions which, on the contrary, constitutes so large a part of romanticism.

It remains a constant with Nietzsche that the 'logical simplicity' of classicism is the best approximation to the hierarchical synthesis achieved by the grand style. The classical style is essentially a representation of calm, simplification, abbreviation and concentration. He makes no mystery of this:

> 'Becoming more beautiful' is a consequence of enhanced strength; it is the expression of a *victorious* will, of increased coordination, of a harmonising of all strong desires. Of an infallibly perpendicular stress and balance. The simplifications of logic and geometry are a necessary consequence of the enhancement of strength. (*The Will to Power*)

We should acknowledge, again, how Nietzsche catches off-guard those who would see him as an enemy of reason, an apostle for the emancipation of the senses from the primacy of logic. He proclaims the opposite, loud and clear: 'We are the adversaries of sentiment and emotion!' The artist worthy of the name is one who cultivates a 'hatred of sentiment, of sensibility, of finer feelings, a hatred for what is inconstant, changeable, vague, superstitious . . .' For, 'to be classical one must possess *all* the strong, seemingly contradictory gifts and desires – but in such a way that they advance together under one yoke' (*The Will to Power*); what is required therefore is 'coldness, lucidity, hardness and logic, above all else.' (*The Will to Power*)

This could not be clearer: classicism is the perfect incarnation of 'the grand style' in morality. Which is why, as against Victor Hugo, whom he takes to be a sentimental romanticist, Nietzsche restores the claims of Corneille, in his eyes a Cartesian rationalist, like one of those

> poets of an aristocratic civilisation . . . who made it a point of honour to *submit their senses*, however vigorous, *to a concept*, and impose upon the brutal claims of colours, sounds and forms the law of a clear and refined intellectuality; in which respect they seem to me to have followed in the steps of the ancient Greeks. (*The Will to Power*)

The triumph of classicism, Greek or French, consists in victoriously combating what Nietzsche again refers to as 'the plebeian sensuality' with which 'modern' – romantic – painters and composers so eagerly fill their works. Contrary to the classical spirit, the romantic hero is usually depicted as someone devoured and therefore diminished by his internal passions. He is unhappy in love; he sighs and weeps; he tears his hair; he leaves the torments of passion only to fall back into those of creation. Which is why, in general, the romantic hero is ill and pale, and invariably dies young, sapped from within by those forces which possess him and undermine him with their failure to harmonise: this is what Nietzsche abhors, and it is why he comes to detest Wagner and Schopenhauer, and why he always prefers Mozart to Brahms – in other words, prefers classical and 'mathematical' to 'romantic and sentimental' music.

Here in fact is an essential aspect of all philosophy, that the practical dimension must join the theoretical, that ethics is not separable from ontology; for in this morality of grandeur it is intensity which counts for most, the will to power which prevails over all other considerations. Which goes to show that there are values, there is an ethics of the immoralist.

Like the disciple of martial arts, the exponent of 'the grand style' moves in a sphere of grace, at a furthest remove from any apparent effort. He does not perspire, and if he moves mountains, he does so with serenity. Just as true knowledge – 'the gay science' – mocks theory and mocks the will to truth, in the name of a different verity, so Nietzsche mocks morality in the name of a different morality. The same is true of the doctrine of salvation.

A New Idea of Salvation

Is it vain to seek a doctrine of salvation in Nietzsche? It is true that doctrines of salvation, of whatever kind, are in his eyes the final expression of nihilism – by which he means the negation of life's here and now in the name of some 'ideal beyond' or hereafter. Mocking the promoters of such doctrines, Nietzsche suggests that, of course, none of them will openly admit to being a nihilist, to preferring extinction to life:

> Of course, one doesn't say never say 'extinction', one says 'the other world', or 'God', or 'the *true* life', or Nirvana, salvation, blessedness . . . This innocent rhetoric, from the realm of religio-ethical balderdash, appears *a good deal less innocent*, however, when one reflects upon the tendency that is concealed beneath these sublime words: the tendency to *destroy life*. (*The Antichrist*, 7)

To find salvation in God, or in whatever figure of transcendence one might wish to put in his place, means to 'declare war on life, on nature, on the will to live! God becomes the formula for every slander upon the "here and now", and for every lie about the "hereafter".' (*The Antichrist*, 18) You can see from these declarations how directly Nietzsche's critique of nihilism confronts the doctrine of salvation, confronts the project of seeking a 'beyond' of whatever variety, an 'ideal' which would 'justify' life, give it a sense, and thereby in some sort save life from the misfortune of being mortal. But does this mean that every impulse towards wisdom or blessedness must, in Nietzsche's eyes, be discarded? Nothing could be further from the truth, since Nietzsche, like every true philosopher, is a seeker after wisdom.

Read the opening chapter of *Ecce Homo*, 'Why I Am So Wise'. This wisdom is progressively entrusted to readers in Nietzsche's late works, and becomes enshrined in his famous – if initially obscure – doctrine of 'eternal recurrence'. This has given rise to so many interpretations and misunderstandings, it is worth our while reviewing its essential outlines.

Recurrence: A Doctrine of Salvation without Gods or Idols

It needs to be said that Nietzsche barely had time to formulate his notion of eternal recurrence before illness prevented him from developing it as fully as he would have wished. Nevertheless, he was wholly convinced that it was in this final doctrine that his most original thought was to be found, his true contribution to the history of ideas.

Its central question is one that concerns us all – or at least all of us who are no longer 'believers'. If there is no longer an elsewhere – a hereafter, a *cosmos*, a divinity – and if the founding ideas of Enlightenment humanism are themselves compromised, how are we to distinguish between good and evil, and (more profoundly), between what is worth living for and what is second-rate? To implement this distinction, do we not need to lift our eyes towards some heaven or other in order to find a transcendent answer here below? And if the sky is hopelessly empty, where shall we turn?

It was to provide a response to this question that Nietzsche formulated the doctrine of eternal recurrence; to afford us quite simply a criterion, of a terrestrial and this-worldly kind, finally, for deciding what is worth living for, and what is not. For those who are believers, this will naturally go unheeded. But for the rest of us, who no longer believe in another world, or for whom this-worldly engagement of whatever kind – political activism, for example – no longer suffices, Nietzsche's answer is worth hearing.

As to whether it corresponds to a doctrine of salvation, or not, of this there can be no doubt. We have only to consider for a moment the manner in which Nietzsche presents his theory, in relation to traditional religion. It offers, he says, 'more than all the religions, which have taught us to despise life as transitory, and to look longingly towards *another* life', so that it will become '*the religion* [par excellence] of the freest and most serene spirits'. Nietzsche goes so far as to propose placing 'the doctrine of eternal recurrence in the place of 'metaphysics' and 'religion' – just as he replaced *theoria* with genealogy, and replaced the ideals of morality with 'the grand style'. We must ask ourselves how he applies them to his own philosophy.

What does the doctrine of eternal recurrence teach us? In what sense does it provide a new answer to the questions of wisdom and of salvation? I suggest a brief answer to these questions. If there is no longer transcendence, or ideals, or possible escape into an elsewhere, however 'humanised' – after the death of God – in the form of moral or political utopias ('humanity', 'fatherland', 'revolution', 'republic', 'socialism' etc.), then it must be at the core of this life on earth that we learn to distinguish between what is worth living for and what must be allowed to perish. It is here and now that we must learn to separate forms of life that are failed – mediocre, reactive, weakened – from forms of life that are intense, grandiose, courageous and rich in diversity.

The first lesson to retain, therefore: that salvation according to Nietzsche cannot be other than resolutely *earthly*, sewn into the tissue of forces that are the fabric of life. Nor can salvation have anything to do with inventing a new ideality, a new idol through which to judge and condemn existence – yet again – in the name of some principle supposedly superior to and exterior to it. All of this is clearly suggested in a crucial text, the Prologue to *Thus Spake Zarathustra*, one of the last things which Nietzsche wrote. In his distinctive style, he invites the reader to the notion of blasphemy upside down:

> I entreat you, my brothers, *remain true to the earth*, and do not believe those who speak to you of extraterrestrial hopes. They are poisoners, whether they know it or not . . . They are destroyers of life, atrophying and self-poisoning, of whom the earth is weary: so let them be gone! . . . To blaspheme against God was formerly the greatest blasphemy. But God died, and his blasphemers died likewise. The most dreadful offence now is to blaspheme the earth, and to prefer interpreting the entrails of the unknowable more than the meanings of the earth.

In a few lines Nietzsche sets down, as no one else had done, what would become in the twentieth century the agenda of every materialist-inspired philosophy; the agenda of all thinking resolutely opposed to 'idealism', understood as a philosophy which would impose ideals superior to the reality that is lived life, or, in Nietzschean terms, our human will to power. Blasphemy here changes its meaning overnight: in the seventeenth and even as late as the eighteenth century, whoever made a public profession of atheism could be thrown in prison, and in some cases put to death. Today, says Nietzsche, the converse should be the rule: to blaspheme is no longer to claim that God is dead, but on the contrary, to succumb yet again to the metaphysical and religious inanities which insist that there is a 'beyond' consisting of higher ideals – however irreligious these might be, such as socialism or communism – in whose name we must 'transform the world'.

Nietzsche explains all of this with great clarity in a fragment dating from 1881, where, in passing, he amuses himself by parodying Kant:

> If, in all that you wish to do, you begin by asking yourself: am I certain that I would wish to do this an infinite number of times? This should be for you the most solid centre of gravity . . . My doctrine says, the task is to live your life in such a way that you *must* wish to live it again – for you will *anyway*! If striving gives you the highest feeling, then strive! If rest gives you the highest feeling, then rest! If fitting in, following and obeying give you the highest feeling, then obey! Only make sure you come to know what gives you the highest feeling, and then spare no means. Eternity is at stake! This doctrine is mild in its treatment of those who do not believe in it. It has neither hell nor threats. But anyone who does not believe merely lives a *fugitive* life in the consciousness of it. (Extract from Nietzsche's 1881 notebook)

(Compare also *The Gay Science*, IV, 31, as well as the celebrated passages in *Zarathustra* where Nietzsche extolls his doctrine, according to which 'all joy [*Lust*] wants eternity'.)

Here, at last, the significance of eternal recurrence becomes clear. It is neither a description of the way of the world nor 'a return to the Ancients', as has foolishly been suggested, not is it yet a prediction. At bottom, it is nothing more than a criterion for deciding which moments in a life are worth living and which are not. Thanks to it, we are enabled to examine our lives so as to avoid pretence and half-measure, all those small acts of weakness, as Nietzsche says again, those concessions to 'just this once', where we give in to the easy exception to any rule, without really wishing to.

Nietzsche invites us to live in such a manner that regrets and remorse have no place and make no sense. Such is the life lived according to truth. Who, after all, would wish that all the instants of mediocrity, the petty struggles, the futile guilt, the hidden weaknesses, the lies, the cowardice, the little arrangements with oneself – that all of this should recur for all eternity? And, by extension, how many instants of our lives would happen in the first place were we to apply, honestly and rigorously, the test of their recurrence? A few moments of joy, no doubt; a few moments of love, of lucidity, of serenity . . .

You might think that this is all very interesting, and possibly useful and true, but appears to have no connection to religious belief, even of a radically new kind, nor to the question of salvation. How exactly can it prevent me from suffering the mortal fear which we examined at the outset of this book? How does it connect to human mortality and its anguish, which the doctrines of salvation attempted to address? Perhaps the notion of eternity will set us on the right track. For you will note that, even in the absence of God, there is *eternity* just the same; and to attain it, we must, Nietzsche insists – strangely, because the assertion seems almost Christian – *have faith and cultivate love*:

> Oh, how not to burn with longing for Eternity and for the wedding-ring of rings – the ring of Recurrence? Never yet have I found the woman by whom I should like to have children, unless it be this woman whom I love: for I love thee, O Eternity! For I love thee, O Eternity! ('The Seven Seals', *Thus Spake Zarathustra*)

The poetry of such passages does not always increase their clarity, I admit. If we wish to understand them and to understand the sense in which Nietzsche revives the doctrines of salvation, we need to realise how close his ideas are to one of the more profound intuitions of ancient philosophy: according to which the good life is that which succeeds in existing for the moment, without reference to past or future, without condemnation or selection, in a state of absolute lightness, and in the finished conviction that there is no difference therefore between the instant and eternity.

Amor fati (Love of What the Present Brings)

We have seen how central this theme was to the Ancients, as to the Buddhists. Nietzsche returns to it in this magnificent passage:

> My formula for greatness in a human being is *amor fati*: to want nothing to be other than as it is, neither in the future, nor in the past, nor in all eternity. Not merely to endure what happens of necessity, still less to hide it from oneself – all idealism is untruthfulness in the face of necessity – but to *love* it . . . (*Ecce Homo*, 'Why I Am So Wise')

Not to wish anything to be other than it is! The maxim could have been written by Epictetus or Marcus Aurelius – whose cosmology Nietzsche never ceased to hold up to ridicule. And yet, Nietzsche clings to it stubbornly, for example in this fragment from *The Will to Power*:

> An *experimental philosophy* such as I live anticipates even the most extreme nihilism . . . But it wants rather to cross over to the opposite – to a *dionysian affirmation* of the world as it is, without subtraction, exception or

> selection. It wants the eternal cycle: the same things, the same logic and illogic of entanglements. The highest state to which a philosopher can attain: to stand in a dionsyian relationship to existence – my formula for this is *amor fati*. Which perceives not merely the *necessity* of those aspects of existence hitherto *denied*, but their *desirability*. (*The Will to Power*, 1041)

To hope a little less, regret a little less, love a little more. Never to loiter in those unreal corridors of time – the past and the future – but try on the contrary to live in and embrace the present as much as possible (with a 'dionysiac affirmation', a reference to Dionysus, the Greek god of wine, festivity and joy – who above all other deities loved life).

Why not? But you may raise a few objections. We can admit, just, that the present moment and eternity resemble each other if neither is 'relativised' and foreshortened by reference to past or future. We can also accept, together with the Stoics and Buddhists, that he who succeeds in living entirely in the present can find in such an attitude the means to escape the anguish of dying. So far, so good. But there remains a troubling contradiction between the two messages which Nietzsche is preaching: on the one hand, in the doctrine of eternal recurrence, he requires us to *choose* what we are live and are willing to relive, given that this will be repeated unavoidably; on the other hand, he urges us to love the real, whatever the case, without picking and choosing, and above all without wishing anything to be other than it is. The doctrine of recurrence invites us to *select* to live only those instants that we would be willing to live with over and over again, in infinite recession – whereas the notion of *amor fati*, which says yes to destiny, makes no exceptions, but comprehends and accepts all of experience within the one perspective: namely, love of the real. How do we reconcile these two positions? By admitting, as far as is possible, that this embrace of destiny kicks in only after the application of the highly selective requirements of eternal recurrence: were we to live under the auspice of eternity, were we finally to discover ourselves in and through 'the grand style', everything that happens to us would be good. The slings and arrows of fortune would no longer have any significance, no more than the happy outcomes. Because we would finally be living reality as a whole, as if each moment were eternity – for a reason that Buddhists and Stoics alike had also grasped: if everything that occurs is necessary, if the real in effect means the present moment, past and future lose their capacity to burden us with guilt, to persuade us that we might act *differently*, and therefore *must* act differently. This explains our attitudes of remorse, of nostalgia, of regret – but equally of doubt and hesitation in respect of the future – which lead to so much inner torment and self-conflict, and therefore to the victory of reaction, since these attitudes inevitably lead our vital forces into mutual confrontation.

The Innocence of Becoming

If the doctrine of eternal recurrence echoes that of *amor fati*, the latter in turn culminates in the ideal of an existence entirely free of guilt. As we have seen, guilt is the essence of what is reactive, the direct outcome of inner torment and self-division. Only the wise man who practises the grand style and lives by the rule of eternal recurrence can attain to true serenity. And this is precisely what Nietzsche means by the expression 'the innocence of becoming': 'to situate oneself beyond every kind of praise and blame, to make oneself independent of everything connected with yesterday and today – so as to pursue my own aim in my own manner'. For it is by this means alone that we can experience salvation. But saved from what? As always: saved from fear. By what means? As always: through serenity. For this reason:

> We who desire to restore innocence to becoming, would like to be the missionaries of a cleaner idea: that no one has given man his qualities, neither God, nor society, nor his parents and ancestors, nor he himself – that no one is to *blame* for him. There is no being who can be held responsible for the fact that someone exists, that someone is thus and thus, that someone was born into certain circumstances, into a certain milieu. – *And it is a tremendous restorative that such a being is lacking* . . . There is no place, no purpose, no meaning, onto which we can shift responsibility for our being, not our manner of being . . . And, to say it once again, this is a tremendous restorative; this constitutes the innocence of all existence. (*The Will to Power*, 765)

Unlike the Stoics, clearly, Nietzsche does not believe that the world is harmonious and rational; the transcendence of the *cosmos* no longer holds. But like the Stoics, he invites us to live inside the moment, to be responsible for our own salvation by accepting everything that is the case, to obliterate in ourselves the distinction between happy and unhappy events, to emancipate ourselves above all from these inner conflicts fatally nurtured by a misunderstanding of time: remorse bound up with a indeterminate vision of the past ('I should have acted differently'), hesitation in the face of the future ('Should I not act differently?'). For it is in freeing ourselves from this insidious double bind of reactive forces (all inner conflict is in essence reactive), in freeing ourselves from the burdens of past and future, that we shall attain to serenity and to eternity, here and now, *because there is nothing else*, no more reference to 'possibility', which would relativise present existence and would sow in us the poisonous weeds of doubt, remorse or hope.

Nietzsche: Criticisms and Interpretations

I have tried to place Nietzsche's thought in the best light, without seeking to criticise. I believe that we must understand before making objections, and that this process takes time, sometimes a lot of time; but also and above all, I believe that we must learn to think with the help of others, and through them to attempt to think for ourselves.

However, there is one specific object – concerning Nietzsche – I must raise, so that you may understand why, despite my considerable interest in the work of Nietzsche, I am unable to be a Nietzschean. This objection concerns the doctrine of *amor fati* which is found in several philosophical traditions, notably Buddhist and Stoic, and also resurfaces in contemporary materialist philosophy (in the next chapter). The notion of *amor fati* sits on these principles: to regret things a little less, to hope for the future a little less and to love the present a little more, if not completely! I can understand perfectly that there can be serenity, relief, solace – everything Nietzsche describes so compellingly – in 'the innocence of becoming'. But this injunction really only applies to the more painful aspects of existence: to enjoin us to love what is already lovable about reality would make little sense, since we do so anyway. What the wise man must manage to realise in himself is the love of whatever happens; otherwise he merely resembles everyone else in liking what is likeable and not liking what is not likeable! And here is the problem: if we must say yes to everything, without 'picking and choosing', but must shoulder whatever comes our way, how do we avoid what one contemporary philosopher and disciple of Nietzsche, Clément Rosset, has so aptly referred to as 'the hangman's argument'. This can be summarised as follows: there exist on Earth, since time immemorial, hangmen and torturers. They are indubitably part of the real; consequently, the doctrine of *amor fati*, which urges us to love whatever is the case, likewise must urge us to love torturers.

Another contemporary philosopher, Theodor Adorno, asked whether, after Auschwitz and the genocide perpetrated against European Jews, mankind could still be urged to love *the real as it is*, without reserve or exception. Is such a thing possible, even? Epictetus, for his part, admitted that he had never in his life met a single Stoic sage, if by this is meant someone who loved the world as it is, under all aspects, however atrocious, and who under all circumstances could refrain from either regret or hope. Must we see in this failure a temporary wobble, a difficulty with the demands of wisdom – or is it not a sign that the theory falters, that *amor fati* is not merely impossible but on occasion obscene? If we must accept everything that occurs, as it is, in all its tragic sense or lack of sense, how can we avoid the accusation of complicity, even of collaboration with evil?

There is more. If loving everything that is the case turns out not to be *truly feasible*, neither for Stoics nor for Buddhists nor for Nietzsche himself, does it not immediately risk taking on the abhorrent form of a new *ideal*, and, consequently, a new figure of nihilism? Here, in my own opinion, is the strongest argument against the long tradition running from the most ancient practices of Oriental and Occidental wisdom to the most up-to-date philosophical materialism. What is the good of pretending to have finished with 'idealism', with all ideals and 'idols', if this proud philosophical programme of *amor fati* remains itself an ideal? What is the good of holding up for derision all theories of transcendence, old and new, and invoking the wisdom of things as they are, if this love of the real is itself in thrall to transcendence and remains an objective that becomes radically inaccessible whenever the going gets even mildly difficult?

Wherever such questions lead, they cannot undermine the historical importance of Nietzsche's responses to the three challenges confronting all philosophy: genealogy as a new *theoria*, the 'grand style' as a new morality, and the innocence of becoming as a doctrine of salvation without God or ideals. These form a coherent whole. In its claim to dismantle the very notion of the ideal, Nietzsche's thought opened the way for the great materialist philosophies of the twentieth century.

I would like to suggest three ways in which the work of Nietzsche has been interpreted. First, we can trace the development of a radical anti-humanism, an unprecedented dismantling of the ideals erected by the Enlightenment. In fact, it is generally accepted that progress, democracy, the rights of man, republican and socialist ideals – all of these idols and more were denounced by Nietzsche, so that when Hitler met Mussolini it was not entirely by chance that he presented him with a handsome bound edition of Nietzsche's complete works. Nor is it an accident that Nietzsche has also served in a context that is different but related – in its hatred of democracy and humanism – namely as a model for the cultural leftism that emerged in the 1960s. We can also see Nietzsche as a paradoxical continuation of Enlightenment philosophy, a progenitor of Voltaire and the French moralists of the eighteenth century. There is nothing absurd about such a thought. In many respects Nietzsche continued the work inaugurated by their critique of religion, of tradition, of the Ancien Regime, and indeed in his tireless exposure of the interests and hypocrisies concealed behind the arras of their great ideals.

Finally, we can read Nietzsche as accompanying the birth of a new world, in which notions of the real and the ideal were replaced by the overriding logic of the will to power. This was to be Heidegger's conclusion, as we shall see in the next chapter, who saw Nietzsche as the 'thinker of technology', the first philosopher to destroy – entirely and without leaving the smallest trace – the notion of 'purposes': the idea that there was a meaning to be sought for in human existence, objectives to pursue, ends to achieve. With 'the grand style', the only remaining criterion by which to define 'the good life', is indeed one of intensity, of force meeting force, to the detriment of all higher ideals. But – once the pleasure of destruction is over – would this not condemn the world to pure cynicism, to the blind laws of the market and unbridled competition?

6
AFTER DECONSTRUCTION: CONTEMPORARY PHILOSOPHY

But first of all, a question for philosophy: once again, why go further? Why not stay with Nietzsche and his corrosive lucidities? Why not rest satisfied, as many have done, with developing his project, with filling the still empty compartments of his thought, and elaborating upon the theses that he has handed down to us? And if we do not like some of them, if we find that his thought flirts uncomfortably with cynicism and with fascist ideologies why not rewind a little, to the Rights of Man, to the idea of the Republic, to the Enlightenment?

These questions cannot be dodged even by the simplest history of philosophy. For to consider the transition from one epoch to another, from one vision of the world to another, is from now on part of philosophy itself. I will state the matter as simply as possible: the deconstruction of the idols of metaphysics revealed too many things for us not to take account of them. It is not possible, even were it desirable, to go backwards. Versions of 'a return' to a prior dispensation never make much sense: if the earlier positions were so feasible and so convincing, they would not have been abandoned, would not have laid themselves open to the rigours of criticism, would never have ceased being in season. The desire to regain lost paradises always proceeds from a lack of historical sense. We can of course try to bring back school uniforms, blackboards and chalk; we may prefer to go back to the Enlightenment, or re-embrace the Republican ideal, but this can never be more than a posture, a performance that ignores time's passage, as if the latter were null and void – which of course is not the case. The problems of advanced democracies are not those of the eighteenth century; our communitarianism has changed, human aspirations have changed, as have our relations with authority and our habits of consumption; new rights and new political actors (ethnic minorities, women, children) have emerged, and there is no point in pretending otherwise.

The same applies to the history of philosophy. Whether we like it or not, Nietzsche asks questions which we cannot pretend have not been asked. We do not think in the same way after him as we did before, as if he had not occurred, as if his famous 'idols' were still standing bolt upright. This is simply not the case. An upheaval has occurred – not only with Nietzsche, but with the whole of what can be called postmodernity: the avant-gardists have passed through, and we can no longer think, write, paint or play music quite as we did before. Poets no longer extol moonlight or sunsets. A certain disenchantment with the world occurred, but was accompanied by new forms of lucidity, and new freedoms. Who today would seriously wish to return to the time of Dickens' *Oliver Twist*, when women lacked the vote, where workers went without holidays, where tiny children laboured, where the countries of Africa and Asia were colonised one after another. Nobody would wish for a return to this, which is why the nostalgia for lost paradises is a display of desire rather than an act of will.

Where does this leave us? And if, Nietzsche is so 'un-ignorable', why not rest there and content ourselves, as have so many of his disciples (Michel Foucault or Gilles Deleuze, among others) with continuing the work of the master? This is one possibility, and we find ourselves today caught between alternatives which might be summarised along these lines: whether to continue along a path opened up by the founders of deconstruction, or to take once more to the high road.

A First Possibility for Contemporary Philosophy

It is of course possible to carry on up the path set out by Nietzsche, or, more generally, that of *deconstruction.* I say 'more generally', because Nietzsche is the greatest but by no means the only 'genealogist', the only 'deconstructor', the only nemesis of idols. There are also Marx and Freud, and since the beginning of the twentieth century, these three have had an extended progeny. And these philosophers of suspicion have been joined by the massed ranks of the human sciences, which have broadly pursued the deconstructive work of the great materialists.

An entire wing of sociology, for example, has undertaken to show how individuals who think of themselves as autonomous agents are in fact entirely determined in their choices, whether ethnic, political, cultural, aesthetic or even sartorial – and by 'class habitus', which is to say, determined by the family and social milieu into which they are born. The hard sciences themselves joined in – starting with biology, which can be used to demonstrate, in Nietzschean mode, that our famous 'idols' are merely a product of the entirely physical functioning of our brains, if not a mere by-product of the necessity of progressive adaptation to its environment by the human species over the course of its history. To take one example, our prejudices in favour of democracy and the rights of man are to be explained, in the final analysis, not by a disinterested intellectual choice, but by the fact that there is more at stake, for our survival as a species, in cooperation and harmony than in conflict and war.

We can continue to think and theorise in the philosophical style inaugurated by Nietzsche, and this essentially has been the path followed by contemporary philosophy. Not that this style speaks with one voice, by any means. It is in fact rich in diversity, and one would be hard pressed to reduce it to the business of pure deconstruction. We should mention, for example, the Anglophone tradition of 'analytical philosophy', which is concerned above all with the functioning of the sciences, and which is regarded by some as all-important, even if it is not much spoken about on the Continent. In another sphere of activity, philosophers such as Jürgens Habermas, Karl-Otto Appel, Karl Popper or John Rawls have attempted, each after their fashion, to pursue the work of Kant, both modifying and extending it to embrace contemporary questions such as social justice and the ethical principles which should regulate discussion between free and equal citizens; or the nature of science and its proper relation to the democratic idea.

In France, and also to a great extent in the United States, it has been deconstruction which has, at least until recent years, prevailed over other currents of thought. As I said, the 'philosophers of suspicion', Marx, Nietzsche and Freud, have had numerous disciples. The names of Althusser, Lacan, Foucault, Deleuze, Derrida and others less known, belong within this configuration, however various their methods. Each has attempted to unmask the idols in which we believe, the concealed and unconscious logic which imprisons us without our knowledge. Following Marx the focus has been upon economic and social relations; following Freud it has been upon language and buried subconscious instincts, and following Nietzsche upon our nihilist tendencies and submission to reactive forces in all their forms.

Where is it going – this interminable trial of the 'idols' of humanism, conducted in the name of lucidity and the critical spirit? What purpose does it serve? And where is deconstruction itself coming from? Beneath its bold and avant-gardist shell, under the claim to be elaborating a 'counter-culture' in order to thwart the ongoing 'idols' of the bourgeoisie, there is the paradoxical risk of making absolutely sacred the real as it is. Which would be entirely logical: by disqualifying these famous 'idols', by refusing to accept that there can be any other horizon of thought than that of 'philosophising with a hammer', we can only end, as Nietzsche does, with his *amor fati*, by prostrating ourselves before the real as it is.

How, in these circumstances, do we avoid the fate of those former revolutionary activist *converts* to the laws of the market-place, who turned into 'cynics' in the most debased sense of the term: disillusioned, shorn of all ambition other than that of an efficient accommodation to the terms and conditions of reality? And must we resign ourselves, in the name of an increasingly problematic notion of lucidity, to paying our last respects to the ghosts of Reason, Liberty, Progress and Humanity? Does nothing remain in these words, which were once so charged with the light of hope, that can escape the rigours of deconstruction and survive demolition?

How to Move Beyond Deconstruction

If deconstruction tips into cynicism and the critique of 'idols' enshrines things as they are, how do we move beyond deconstruction? For me, these questions open up another path for contemporary philosophy. Not that of a turning back towards Enlightenment, reason, the republic, humanism – which would make no sense – rather, an attempt to rethink them, not 'as before' but after deconstruction, and in its light.

Not to make such an attempt is to risk submitting to things as they are. In which case deconstruction, which set out to liberate minds and break the chains of tradition, has involuntarily turned into its opposite – a new form of adaptation, disillusioned rather than clear-sighted, to the hard reality of a globalised universe. We cannot hedge our bets for ever; on the one hand advocating with Nietzsche *amor fati* and good riddance to all 'higher ideals', but at the same time shedding crocodile tears for the disappearance of utopian aspirations and the harshness of a rampant capitalism.

To become fully aware of this predicament, I need to enlist the thought of Martin Heidegger, who remains in my view the most important contemporary philosopher. He too was one of the founding fathers of deconstruction, but his thought is not a version of materialism and is not hostile to the idea of the transcendental. He is to my mind the first to have given the contemporary world – what he refers to as 'the world of technology' – a reason why we cannot remain content with being Nietzscheans, if we do not wish to become complicit with a reality which today takes the form of capitalist globalisation. Despite its extraordinarily positive aspects – formidable economic growth and wider distribution of wealth than ever before – it also has devastating effects upon the life of the mind, on the political sphere and, fundamentally, on our existence.

By way of an introduction to contemporary philosophy, I would like to begin by exploring this fundamental aspect of Heidegger's thinking. First of all, because it is an inspired and brilliant body of thought, and one which sheds incomparable light upon the present. Second, because it equips us to understand not only the economic, cultural and political landscape which surrounds us, but also to grasp why the tireless pursuit of Nietzschean deconstruction can at this point lead only to the hallowing of the realities – however trivial and un-sacred they may be – of a liberal universe given over to accommodation.

Many people say as much today, ecologists in particular, or those who describe themselves as 'alter-globalists'. But the originality of Heidegger and his critique of 'the world of technology' is that it does not content itself with the habitual criticism of capitalism and liberalism. Usually, the latter are reproached indiscriminately for increasing inequality, destroying regional cultures and identities, reducing biological diversity and the species-count, widening the gap between rich and poor, and so on. All of which is not only highly questionable but misses the essential. (It does not follow, for example, that poverty increases in the world if inequality widens, nor that rich countries are unconcerned about the environment. On the contrary, developed societies are infinitely more concerned than poor countries – for whom the necessities of development take precedence over those of conservation – just as they are also the first to see public opinion become truly preoccupied with the preservation of local identities and cultures.)

All of the above can be debated at length, but what is certain and what Heidegger enables us to understand, is that liberal globalisation is in the process of betraying one of the most fundamental promises of democracy – how collectively to make our own history, to participate in it and have our say about our destiny, and to try and change it for the better – because the world which we are entering not only 'escapes' us on all sides, but turns out to be devoid of sense: stripped of meaning and of direction.

Each year, your mobile phone, your MP3 player and computer games change, along with everything else around you: their functions multiply, they become smaller, their screens get bigger or become coloured, and so on. And you know that a product which does not keep in step is going to fail. Unless it follows suit. It is not a question of taste, of one choice among others, but a necessity without choice, in which survival is at stake.

In this sense, we could say that in today's world of globalised capital which places all human activities in a state of perpetual and unending competition, history is moving beyond the will of men. Competition is becoming not only a form of destiny, but, what is more, there is nothing to suggest that it is moving in the direction of what is better. Who can seriously believe that we shall have more freedom and be happier because in a few months the weight of our MP3 players will have halved, or their memory doubled? In accordance with Nietzsche's wishes, the idols are all dead: no ideal, in effect, animates or disturbs the course of things, only the absolute imperative of change for the sake of change.

To use an ordinary but suggestive image: as a bicycle must keep going in order not to topple over, or a gyroscope must keep spinning to remain on its axis, we must ceaselessly 'progress'; but this mechanical progress induced by a struggle for survival can no longer be integrated within a grand design. Here too, the transcendental bias of the great humanist ideals Nietzsche mocked has well and truly disappeared – so that in a sense it is indeed Nietzsche's programme that has been accomplished to perfection by globalised capitalism – as Heidegger suggested was the case.

The difficulty is not so much that globalisation supposedly impoverishes the poor in order to engorge the rich, as ecologists and alter-globalists suggest, but that it dispossesses us all of any purchase on history, and divests history itself of all purpose. Dispossession and directionlessness are the terms which best characterise it – in which respect again it fulfils perfectly, in Heidegger's eyes, the philosophy of Nietzsche: a body of thought which assumed, as no other has ever done, the complete eradication of all ideals at the same time as the logic of historical direction.

The Advent of a 'World of Technology' and the Retreat of Meaning

In a brief essay entitled *Overcoming Metaphysics*, Heidegger described the domination of technology which characterises the contemporary universe as the result of a process which took root in seventeenth-century science and spread slowly into all areas of democratic life.

I would like to offer the principle aspects of this argument in simple language, for those who have not yet read any Heidegger. I should warn you, however: what I am going to say will not be found in this form in Heidegger. I have added various examples which are not his, and I present his technical argument in my own non-technical fashion. Nevertheless, the central idea is certainly his, and what matters here is not the provenance of any particular concept, but the idea to be drawn from Heideggerian analysis: according to which, the project of mastery of nature and history which accompanied the birth of the modern world and which gives all its meaning to the democratic idea, can be seen to turn finally into its exact opposite. Democracy promised us the possibility of taking part in the collective construction of a free and fair world. Yet today we are losing almost all control over the course of the world in which we live – a supreme betrayal of the promises of humanism, implicit in democracy, and one which raises many questions.

The first moment of the process Heidegger describes coincides with the birth of modern science, which broke with ancient philosophy at every point and which saw the emergence of a project of domination over Earth, of its total mastery by the human species. According to the famous formula of Descartes, scientific knowledge would permit man to make himself 'as if the master and owner of nature': 'as if', because he was not yet entirely alike to God, his creator, but almost. This aspiration to scientific domination takes a dual form.

It was to express itself first on a straightforwardly 'intellectual' or theoretical level: that of knowledge about the world. Modern physics was entirely founded on the premise that nothing, in the world, occurs without reason. In other words, everything must be rationally explicable, sooner or later; every event has its cause, a reason for being, and the role of science is to discover these reasons. Scientific progress became merged with the progressive eradication of the mystery that, in the Middle Ages, was believed to be part and parcel of nature.

A second impulse to domination emerges behind that of the need for knowledge; this time an entirely practical dominance, proceeding not from the intellect but from the will of men. If nature is no longer mysterious or sacred but on the contrary can be reduced to an inventory of merely physical phenomena entirely devoid of meaning or value, then there is nothing to prevent us from harnessing nature in whichever way seems appropriate for our ends. To take an example, if the tree growing in the forest is no longer (as it was in the fairytales of our childhood) a magical being, likely to transform itself into a witch or a goblin during the night, but merely a piece of wood devoid of a soul, nothing prevents us from turning it into furniture or chucking it on the fire to warm ourselves. Nature as a whole loses its spell, and becomes a vast warehouse on which humans can draw at will, without restriction other than that imposed by a conceivable need to keep something back for the future.

For all that, with the birth of modern science we have still not quite arrived at what Heidegger would call 'the world of technology', which is to say a universe in which the preoccupation with ends – with the ultimate purpose of human history – has totally disappeared, in the interests of an overriding and exclusive preoccupation with means. In seventeenth- and eighteenth-century rationalism – in the thought of Descartes, or the Encyclopedists, or Kant, for example – the project of a scientific mastery of the universe still possesses an emancipatory purpose, by which I mean that, in its principles, it remains subject to the fulfilment of certain ends and objectives considered to be beneficial for humanity. We are not as yet exclusively interested in the means which will enable us to dominate the world, but in the objectives which such domination might enable us to realise. In this respect, clearly, human interest in domination has not yet become purely technological. If this means dominating the universe both theoretically and practically, through scientific knowledge and the exercise of will, it is not merely for the pleasure of domination or a fascination with our own powers. The project is not about mastery for mastery's sake, but about understanding the world and, if necessary, being able to exploit it in order to reach certain higher objectives, which ultimately can be grouped under two headings: liberty and happiness. In this sense, it becomes clear that at its birth modern science had not yet been reduced to pure technology.

From Science to Technology: the Disappearance of Ends and the Triumph of Means

During the seventeenth and eighteenth centuries, science still rested on two convictions which underpinned an Enlightenment optimism in human progress. The first conviction is that science will allow us to liberate our spirit, to emancipate humanity from the shackles of superstition and medieval opposition to new knowledge. Reason will emerge triumphant from its combat with religion and, more generally, from the combat against all forms of argument based on the monopoly of authority. And in this sense, as we have seen in connection with Descartes, modern rationalism prepared the way intellectually for the French Revolution. The second conviction is that mastery of nature will liberate us from the ills and natural servitude to which we are heir, and turn them to our advantage. You will recall the emotions provoked by the famous Lisbon earthquake of 1755 which, in a matter of hours, killed thousands of people, and set in train a debate between philosophers as to the 'wickedness' of a natural order which bore no relation to received ideas about a harmonious and well-intentioned *cosmos*. Virtually everyone at the time concluded that science would save us from the tyranny of nature. Thanks to science, it would be possible finally to foresee and therefore prevent the catastrophes so regularly visited upon us by nature. Here the seeds of the modern idea of happiness dispensed by science, of wellbeing made possible by mastery over the world, make their first appearance.

And it is on account of these two convictions or purposes – freedom and happiness, which together define the idea of progress – that the development of the sciences appears as the vector of another idea, that of civilisation and its march. No matter that such a vision strikes us as naïve or otherwise. What counts is that the will to mastery over nature is still linked to higher objectives and motives, and in this sense cannot be reduced to a purely instrumental or technological rationalism.

For this vision of the world to become thoroughly technological required only one more step; that the project of the Enlightenment be integrated and 'docked' with the world of competition, so that the engine of history – the evolutionary principle – ceased to be linked to any vision or ideal, to become instead the mere outcome of competition.

The Passage from Science to Technology: The Death of the Great Ideas

In this new perspective, that of generalised competition – which we refer to today as 'globalisation' – the idea of progress changes its meaning completely: instead of being inspired by transcendental ideals, the progress of society (or, more neutrally, its forward *movement*) is gradually reduced to meaning no more than the automatic outcome of the free competition between its constituent parts.

At the core of businesses, but also of scientific laboratories and research centres, the unceasing imperative to measure oneself against others (what is today known by the awful term 'benchmarking'), to increase productivity, to develop expertise and above all to apply the fruits to industry and the economy – consumption, in other words – has become an absolutely vital imperative. The modern economy functions like Darwinian natural selection: within the logic of globalised competition, a business which does not 'progress' each day is quite simply doomed to extinction. But this advance has no other end than itself – to stay in the race with the other competitors.

Hence the fearsome and incessant development of technology, tethered to and largely financed by economic growth, and the fact that the increase of human power over nature has become completely automatic, uncontrollable and blind, because it everywhere exceeds the conscious will of the individual. And this is, quite simply, the inevitable result of competition. In which sense, contrary to the philosophy of the Enlightenment, which aimed at emancipation and human happiness, technology is well and truly a process without purpose, devoid of any objectives: ultimately, nobody knows any longer the direction in which the world is moving, because it is automatically governed by competition and in no sense directed by the conscious will of men collectively united behind a project, at the heart of a society which, as recently as the last century, could still think of itself as *res publica*: 'the common weal'.

In the technological world, which from now on means the world as such, since technology is a planetary phenomenon without limits, it is no longer a question of dominating nature or society in order to be more free or more happy, but of mastery for mastery's sake, of domination for the sake of domination. Why? For no end, precisely, or rather: because it is quite simply impossible to do otherwise, given the nature of societies entirely governed by competition, by the absolute imperative to 'advance or perish'.

Now we understand why Heidegger calls the universe in which we live 'the world of technology', or the technical world, and let us think for a moment of the significance assumed by the word 'technique' in current usage. It generally designates the ensemble of means required in order to achieve a given end. It is in this sense, for example, that we say of a painter or pianist that he or she possesses a 'good technique', to indicate that they master their art sufficiently to be able to paint or play whatever they wish. It is important to note that, before all else, technique concerns means and not ends, that is, it can be placed in the service of different ends, but does not of itself choose them: essentially the same technique will serve a pianist playing classical or jazz, traditional or modern, but the question of choosing which works to play does not in any sense derive from technical competence. The latter operates in a world of 'if this . . . then that'. 'To achieve this, you must do that', it says – but never does it tell us what to choose as an objective, or why. A 'good doctor', in the sense of a good technician of medicine, can both kill or cure his patient – the first perhaps more easily than the second. But the decision to kill or to cure is indifferent as far as the logic of technique is concerned.

It is equally legitimate to say that the universe of globalised competition is, in a broader sense, 'technical'. For there too, scientific advancement well and truly ceases to have ends in view that are exterior to or higher *than* itself, but becomes a kind of end in itself – as if the proliferation of means, of power and mastery of mankind over nature have become their own finality. It is precisely this 'technicalisation of the world' which occurs in the history of thought, according to Heidegger, with the Nietzschean doctrine of 'the will to power', which deconstructed and even destroyed all 'idols', all higher ideals. In reality – and no longer merely in the history of ideas – this mutation occurs in the advent of a world where 'progress' has become a process automated and divested of purpose, a sort of self-regulating mechanism from which human beings are totally dispossessed. And it is just this disappearance of ends in the interests of an overriding logic of means that constitutes the victory of technology.

Here is the final difference, the gulf separating us from the Enlightenment and dividing the contemporary world from that of the Moderns: no one can be reasonably convinced any longer that this teeming and disruptive evolutionary impulse, this incessant movement unconnected by a common project, leads infallibly towards what is better. Ecologists are sceptical, as are critics of globalisation, but equally republicans and even liberals become nostalgic for a time that is still recent yet seems irrevocably of the past. From which comes a sense of doubt. For the first time in the history of life, a living species holds the means to destroy the entire planet, and this species does not know where it is going. Its powers of transformation and, if need be, of destruction, are by now unbounded, but like a giant with the faculties of an infant, they are totally dissociated from any capacity for reflection – while at the same time philosophy itself withdraws from engagement with such questions, likewise seized by a passion for the technical.

No one today can seriously claim to believe in a guarantee of survival for our species, and many are troubled, but no one knows how to take the reins: from Kyoto protocols to ecological summits, our heads of state participate impotently, brandishing a rhetoric crammed with pious hopes but with no real power to control even those scenarios most clearly identified as potentially catastrophic. The worst does not always come to pass, and nothing prevents us of course from remaining optimistic, but this is more an act of faith than having any basis in reason. Hence the Enlightenment ideal gives way to a diffuse and multiform anxiety, always at the ready to focus upon this or that particular threat, in such a way that fear is slowly becoming the characteristic democratic emotion.

What lessons are to be drawn from such an analysis? First, that the genealogical and technical attitudes are, as Heidegger thought, two sides of the same coin: the first is the close-fitting philosophical double of the second, which is merely its social, economic or political equivalent. There is a paradox here, of course. On the surface, nothing could seem further removed from the technical world – with its democratic mandate, insipid and collectivist, at the opposite pole to any notion of a 'grand style' – than the aristocratic and poetic formulations of Nietzsche. However, by smashing all our idols with his hammer, and delivering us – in the guise of clear-sightedness – bound and gagged to the world of whatever is the case, Nietzsche's thought serves however unintentionally the incessant flux, the hither and thither of modern capitalism.

From this point of view Heidegger is correct, and Nietzsche is well and truly 'the thinker of technology' – the philosopher who, like no other, sings the disenchantment of the world, the eclipse of meaning, the disappearance of higher ideals in the interest of the single-minded logic of the will to power. That Nietzsche was held up as something of a radical utopian during the 1960s is one of the great blunders in the history of misinterpretation. Nietzsche is an avant-gardist, of course, but certainly not a thinker of utopias. Quite the opposite, he is their most ardent and effective denigrator.

The risk is therefore great that the indefinitely prolonged and inexhaustible pursuit of deconstruction will only be laying siege to a door that is already fairly wide open. The problem is no longer, regrettably, that of breaking yet again the poor 'clay feet' of those unfortunate ideals that no one can manage any longer even to identify, so fragile and uncertain have they become. The urgent need is certainly no longer to challenge concealed 'powers' by now so invisible as to be non-existent, so mechanised and anonymous has the course of history become; but on the contrary, to enable new ideas and even new ideals to arise vigorously, so as to regain a minimum of control over the shape of things. For the real problem is not that history is in the covert hands of figures of 'authority', but on the contrary that it now eludes all of us, authorities included. It is no longer power that inhibits us, but the absence of power – so that the desire to keep on deconstructing idols, to keep on discovering yet again the hiding places of 'Power' with a capital 'P', is not so much to work for the emancipation of mankind as involuntarily to conspire with a blind and demented globalisation.

The priority, in our current situation, is, as we have said, to take the reins: to attempt if possible to 'keep our mastery in check'. For his part, Heidegger did not believe that this could be done, or rather did not believe that democracy was equipped for such a challenge – which is no doubt one of the reasons why he embraced instead the worst authoritarian regime mankind has ever known. He believed, in effect, that democracies are fatally wedded to the structures of the technical world. Economically so, because they are intimately bound to the liberal creed of competition. And this system, as we have seen, of necessity triggers an unlimited and automatic proliferation of productive forces. Politically, likewise, because elections also take the form of organised competition which imperceptibly tends towards a logic whose fundamental elements – that of the popular vote and the supremacy of the ratings poll – are the very essence of the technical world, the society of globalised competition.

Heidegger therefore chose Nazism, convinced without any doubt that only an authoritarian regime could prove equal to the challenge to humanity posed by the technical world. Subsequently, in his later writings, he distanced himself from all voluntarism, all attempts to transform the world. Although understandable, either of these positions is unpardonable, even absurd – which proves that tragically mistaken conclusions may be drawn from an analysis of genius. A large part of Heidegger's work is for this reason desperately disappointing, and sometimes unbearable, although the essence of his conception of the technical world is extraordinarily illuminating.

Two Possible Avenues for Contemporary Philosophy

In the technical world, philosophy can take two very different directions. First, we can make of philosophy a new 'scholasticism', in the proper sense of the term: a discipline confined to school and university. The fact is that after an intense historical phase of 'deconstruction', inaugurated by Nietzsche's hammer and pursued under diverse guises, philosophy, itself in thrall to technique, divided itself into specialised categories: philosophy of science, of logic, of law, of morals, of politics, of language, of environment, of religion, of bioethics, of the history of ideas (Occidental and Oriental), Continental or Anglophone philosophy, philosophy by historical periods, by country . . . In truth, there is no end to the 'specialisms' which students are required to choose in order to be considered 'serious' and 'technically competent'.

In research organisations, young people who do not work on a rigorously specialised subject – 'the brains of leeches', as Nietzsche already joked – have no chance of being considered true researchers. Not only is philosophy required to ape at all cost the model of the 'hard' sciences, but these hard sciences have themselves become 'techno-sciences', in other words often more preoccupied by the economic or commercial spin-offs of their activity than by fundamental questions.

When university philosophy wants to take the broader view – when for example a philosopher is asked to pronounce as an 'expert' upon this or that question concerning society – it maintains that its true role is to diffuse a critical and 'enlightened' spirit on questions which it has not raised of its own accord, but which are of general interest. According to which the highest purpose for philosophy would ultimately be a moral purpose: to clarify public debate, promote rational argument in the hope that by so doing we will keep moving in the right direction. And to arrive at this point, philosophy believes, out of intellectual probity, that it must specialise in very particular areas – subjects in which the philosopher (a *professor of* philosophy), ends by acquiring a particular competence.

Today, many universities throughout the world interest themselves in bioethics or ecology, with the aim of studying the impact of positivist science upon the evolution of human societies, so as to furnish answers as to what it is advisable to do or not to, to authorise or prohibit, on topics such as cloning, genetically modified organisms, eugenics or medically assisted procreation. Clearly there is nothing unworthy about such a role for philosophy. Quite the contrary, it can have its uses, and I would not dream of denying these. It is nonetheless dreadfully reductive, when one thinks of the ideals which were common to all the great philosophers from Plato to Nietzsche; none of whom had renounced to this extent their responsibility for pondering what is meant by the good life – or persuaded themselves that critical reflection and moral pronouncement were the ultimate horizons of philosophy.

Faced with this development – not in my view to be confused with progress – the great philosophical questions can seem like the sentimental films of yesteryear, at least as far as these new specialists in seriousness are concerned. No more discussion of meaning, or of what constitutes doing and living well, or the nature of wisdom, or (even less) of salvation! Everything that for several millennia had constituted the essence of philosophy would seem to have been written off, to leave room only for erudition, for 'reflection' and 'the critical spirit'. Not that these attributes are not qualities, but in the end, as Hegel said, 'erudition begins with ideas and ends with ordures'. Everything and anything can become an object of erudition, bottle tops as much as concepts, so that technical specialisation produces forms of expertise that are closely allied to the most arid absence of meaning.

As for 'critical reflection', I have already had occasion to explain, in the opening pages of this book, what I think of this indispensable faculty: that it is an essential requirement in our egalitarian world, but that in no respect whatsoever is it the prerogative of philosophy. Every human being worthy of the name reflects upon his work, his love life, the newspapers, politics and the places in which he finds himself, without thereby becoming a philosopher.

This is why some of us today prefer to set up shop at a discreet distance from the great avenues of academic thought, and likewise from the diagonal crossways of deconstruction. And we would wish, not to restore the old questions – as I have suggested, dreams of 'return' never make sense – but to revisit them, so as to rethink them afresh. It is in this perspective that authentically philosophical discussion remains alive. After deconstruction, and to one side of empty erudition, philosophy sets off once more towards different horizons – more promising ones, in my view. I am convinced that philosophy can and must – more than ever before, given the technical universe in which we are immersed – keep alive the philosophical questions, not only concerning *theoria* and ethics, but also the question of salvation, even if this means renewing the latter from top to bottom.

We can no longer be content today with a philosophical practice that is reduced to the status of a specialised university discipline, nor stick to the logic of deconstruction alone, as if its corrosive clarity were an end in itself. Erudition stripped of meaning is not enough, and the much vaunted critical spirit, even when it serves the ideals of democracy, is merely a necessary but not a sufficient condition of philosophy: it enables us to shake off the illusions and innocence of classical metaphysics, but does not in any sense offer a response to the existential questions which the search for wisdom inherent in the very notion of philosophy used formerly to place at the core of its doctrines of salvation.

We can of course turn our backs on philosophy. We can proclaim loudly and clearly that it is dead, finished, definitively ousted by the human sciences. But we cannot remain content with the dynamic of deconstruction alone and by making an impasse out of the notion of salvation (in whatever sense we intend it). And if we prefer not to yield to the cynicism of *amor fati* we must try to go beyond philosophical materialism. In other words, for whoever is not a believer, for whoever refuses to content themselves with fantasies of a return to a golden age, or confine themselves to philosophising with a hammer, it is necessary to take up the challenge of a wisdom or a spirituality that is post-Nietzschean.

Such a project supposes keeping one's distance from contemporary materialism, of course, with its rejection of all transcendental ideals and their relegation by genealogy to nothing more than the illusory by-products of history and nature. We must show how materialism, even at its most persuasive, does not answer satisfactorily the questions of knowledge or spirituality. I would like to explain this in more detail before suggesting how a post-Nietzschean humanism can succeed in rethinking *theoria*, morality and the problematic question of salvation – or what might stand in its place.

The Failure of Materialism

Even when it chooses openly to address ethics, or indeed a doctrine of salvation – which Nietzsche, for example, only ever attempted to do surreptitiously – contemporary materialism seems unable to command sufficient coherence to be persuasive. Which does not mean that there is no truth in the materialist position, nor compelling intellectual principles, but rather that the attempts to have done with humanism end in failure.

I would like to dwell for a moment on this renewal of materialism – which has similarities with Stoicism, with Buddhism and of course with the thought of Nietzsche – because through its failure, indirectly, as I have suggested, a new humanism may be conceivable.

In the context of contemporary thought, the French philosopher André Comte-Sponville has probably pushed furthest and with most rigour the attempt to found a new ethics and a new doctrine of salvation, on the basis of a radical deconstruction of the claims of humanism to transcendency of ideals. In this sense, even if Comte-Sponville is no Nietzschean – and strenuously rejects the fascistic overtones to which his philosophy is sometimes prey – he shares with Nietzsche nonetheless the conviction that the 'idols' of authority are clay-footed, that they need to be deconstructed, traced back genealogically to their origins, and that the only possible wisdom is one of radical (this-worldly) immanence. His thought, too, culminates therefore in one of the numerous versions of *amor fati*, in an appeal for reconciliation with the world as it is, or, if you prefer, in a radical critique of hope. 'Hope a little less, love a little more' – such is, at bottom, the key to salvation for Comte-Sponville. For hope, contrary to what is commonly thought, far from helping us to live better, in effect makes us forego the essence of life, which is here and now. As for Nietzsche and the Stoics, hope is manifestly a misfortune rather than a virtue. In a maxim as encompassing as it is terse Comte-Sponville summarises: 'To hope is to desire without consummation, without knowledge, without power.' It is therefore a blight, and not an attitude which can give any zest to life.

This maxim describes hope, first and foremost, as to desire without consummation, for by definition we do not possess the objects of our hopes. To wish to be richer, younger, healthier and so on, is necessarily not to be in possession of these things. It is to place ourselves in a relation of absence to what we would wish to have or be. But hope also means to desire without knowledge: if we knew when and how the objects of our hopes were to be realised, we would no doubt content ourselves with waiting for them – and waiting has a different meaning to hoping. Finally, hope means to desire without power since, self-evidently, if we had the capacity or the power to act out our wishes, here and now, we would not deprive ourselves but would put them into effect without the preliminaries of hoping for them.

This reasoning is faultless. Frustration and impotence are the salient properties of hope, from a materialist point of view – in which respect Comte-Sponville's critique shares a kinship with the spirituality of Stoicism and Buddhism. From Greek wisdom the materialist doctrine of salvation freely borrows the famous notion of *carpe diem* – 'seize the day' – that is, the conviction that the only life worth the pain of living is located in the here and now, in our reconciliation with the present.

According to which the two evils which ruin human existence are nostalgia for a past which no longer exists and the expectation of a future which has yet to exist; thus, we miss life as it is, the only life with any validity: a present moment which we must finally learn to embrace for what it is. As with the Stoic message, but also as with Spinoza and Nietzsche, we must endeavour to love the world, and ascend to the level of *amor fati* – this being the final word on the subject from what we might call, paradoxical though it may sound, a materialist 'spirituality'.

Nor can its invitation leave us entirely cold. I am convinced that it has its own truth, which corresponds to an experience we have all had: those moments of 'grace' in our lives, when, by good fortune, the world as-it-is ceases to seem threatening, vile or ugly, but on the contrary benevolent and harmonious. This might arise through a walk along a riverbank, the natural beauty of a landscape, or even – within society – when a conversation or an encounter overwhelms us. All of these examples I have borrowed from Rousseau. Each of us can remember for ourselves such moments of weightless happiness, when we experience a sense that the real is in no need of transformation or improvement, through our hard work, but is there to be savoured for the sake of the moment, without past or future – in joy and contemplation rather than in the hope of better days.

It is clear that, in this sense, materialism is a philosophy of happiness; and when all is going well, who would not willingly yield to its charms? A philosophy for the good times, in short, but can we still follow its lead when the weather turns nasty? This is precisely where our materialist guide might be of some assistance, but suddenly he slips from our grasp – which is what the greatest philosophers, from Epictetus to Spinoza, have been forced to concede: the true sage is not of this world, and beatitude remains, sadly, inaccessible. Faced with imminent catastrophe – a sick child, the rise of fascism, an urgent political or military decision – I know of no materialist sage who does not instantly turn into a vulgar humanist, weighing up the alternatives, suddenly convinced that the course of events must in some sense depend on his free choices.

That we must prepare for misfortune, even anticipate it, as has been said, in the mood of the future perfect tense ('When it comes, I will at least have been prepared for it'), I wholeheartedly agree. But that we must embrace what happens under whatever circumstances seems to me quite simply impossible. What meaning can the imperative of *amor fati* have confronted with the fact of Auschwitz? And what value can our revolutions and our acts of resistance have if they are inscribed for all eternity in the real, alongside and undifferentiated from everything to which they are opposed? I have yet to encounter a materialist, ancient or modern, who was able to provide an answer to this question. Which is why, all things considered, I prefer to commit myself to the path of a humanism which has the courage fully to take on the problem of transcendence. For this is what is at stake: our logical incapacity to put aside the notion of liberty as we have encountered it in Rousseau and Kant – the idea, in other words, that there is within us something in excess of nature or history.

Contrary to what is claimed by materialism, we are unable to think of ourselves as totally determined by history and nature; we are unable to eradicate the sense that we can detach ourselves sufficiently to be able to look upon them critically. One can be a woman and yet refuse to be determined by what nature appears to have planned in the matter of womanhood: child-rearing, family life; one can be born into a socially disadvantaged milieu and yet transform oneself, thanks to education perhaps, and enter worlds quite different to those which a social determinism would have programmed for us.

To convince yourself of this, reflect for a moment upon a logic which you have undoubtedly experienced, which we all experience whenever we make a value judgement. To take a particular example: like more or less everyone, you cannot help but feel that the Bosnian Serb armed forces which ordered the massacre of Bosnian Muslims at Srebrenica were wicked. Before the slaughter, they amused themselves by terrifying their victims, shooting them in the legs, making them run before mowing them down, cutting off their ears, torturing and then murdering them. I cannot see how one can think about the perpetrators of these acts other than as wicked. When I say this, it is self-evidently because I presume that, like other human beings, these men could have acted differently; they possessed *freedom of choice*. If the Serb generals responsible for these acts of genocide were bears or wolves, I would not think of bringing a value judgement. I would merely deplore the massacre of innocent men by wild beasts, but it would not occur to me to judge from a moral point of view. If I do so, it is precisely because generals are not wild beasts but human beings, to whom I attribute the capacity to choose between alternatives.

From a materialist perspective, one might of course argue that such value judgements are illusory. One could trace their 'genealogy', show their direction of origin and bias, how they are determined by a particular history, milieu, education and so on. The problem, however, is that I have yet to meet anyone, materialist or otherwise, who was able to dispense with value judgements. On the contrary, the literature of materialism is peculiarly marked by its wholesale profusion of denunciations of all sorts. Starting with Marx and Nietzsche, materialists have never been able to refrain from passing continuous moral judgement on all and sundry, which their whole philosophy might be expected to discourage them from doing. Why? Quite simply because, without realising it, they continue to attribute to human beings a freedom in everyday life which they deny to them in philosophical argument – to such a degree that we can only conclude that the illusion resides less in the idea of liberty than in the theses of materialism itself, which quite simply prove to be unsustainable.

Beyond the moral sphere, all judgements of value – from a remark about a film you have enjoyed to some music that has affected you – implies that you believe yourself to be free, that you represent yourself as speaking freely rather than as a being in the grip of unconscious forces which talk across you, so to speak, without your being aware of the fact.

What must we trust, then? Your own sense of yourself as acting freely, which is implicitly the case whenever you utter a judgement? Or the materialist, who tells you (freely?) that you are nothing of the kind – while himself continuing to scatter value judgements whenever the occasion arises, all of which presuppose his own freedom? It's your choice, so to speak.

For my part, I would prefer if possible not to exist in a continuous state of self-contradiction. To which end, I attribute to myself – even if it remains a mysterious business, like life – a faculty of self-removal from both nature and history: the faculty which Rousseau and Kant called liberty or perfectibility, even if this occupies a position of transcendence in respect of the historical and biological codes within which materialism would imprison us. I would add for good measure, and to explain the simple reflex of value-judgement which I have been describing, that there exists not merely a transcendence of liberty, so to speak, *within* us, but also values which reside *outside us*: that it is not we who invent the values which guide and animate us; not we who invent, for example, the beauty of nature or the power of love.

Let me make myself clear: I am not saying that we 'need' transcendence, as a somewhat inane modern habit of thought is given to proclaiming (that we 'need' meaning, that we 'need' God). Such formulas are problematic, because they instantly rebound on those who use them: it is not our need for something that proves its existence. Quite the contrary: there is a strong likelihood that the need pushes us to invent the thing, and then to defend it, with all the arguments of bad faith at our disposal, because we have become attached to it. The need for God is, in this respect, the greatest argument against His existence that I know of.

I am not saying, therefore, that we 'need' the transcendence of being free agents, or the transcendence of values. I am saying that we cannot dispense with them, which is quite a different matter; that we cannot think about ourselves, or our relation to values, without positing the hypothesis of transcendence. It is a logical necessity, a rational constraint, not an aspiration or a desire. What is being debated here is not a matter of our comfort, but of our relation to truth. Or, to put the matter differently: if I am not convinced by materialism, it is not because it seems uncomfortable, or lacking in solace. Quite the contrary. As Nietzsche said, moreover, the doctrine of *amor fati* is a source of solace like none other, the ground of an infinite serenity. If I feel obliged to go beyond materialism, to try and push things further, it is because I find it literally 'unthinkable' – too full of logical contradictions for me to settle down with intellectually.

To outline once again the principal ground of these contradictions, I will say that the cross of materialism is that it never quite succeeds in believing what it preaches, in thinking its own thought. This may sound complicated, but is in fact simple: the materialist says, for example, that we are not free, though he is convinced, of course, that he asserts this freely, that no one is forcing him to state this view of the matter – neither parents, nor social milieu, nor biological inheritance. He says that we are wholly determined by our history, but he never stops urging us to free ourselves, to change our destiny, to revolt where possible! He says that we must love the world as it is, turning our backs on past and future so as to live in the present, but he never stops trying, like you or me, when the present weighs upon us, to change it in the hope of a better world. In brief, the materialist sets forth philosophical theses that are profound, but always for you and me, never for himself. Always, he reintroduces transcendence – liberty, a vision for society, the ideal – because in truth he cannot *not* believe himself to be free, and therefore answerable to values higher than nature and history.

From which arises the fundamental question for contemporary humanism: how to formulate transcendence under both aspects – within ourselves (as liberty) and outside ourselves (as values) – without falling immediately back into the clutches of a materialist genealogy and a materialist deconstruction. Or: how to formulate a humanism which is finally relieved of those metaphysical illusions which it was carting around with it right from the start, at the birth of modern philosophy.

Towards a New Idea of Transcendence

Contrary to materialism, to which it is diametrically opposed, the post-Nietzschean humanism of which I dream in these pages – a long tradition which has its roots in the thought of Kant and flowers in that of one of his greatest twentieth-century disciples, Husserl – rehabilitates the idea of transcendence. But it also affords it, notably on a theoretical level, a new meaning which I would like to try and explain here. For it is through this new emphasis that it manages to escape the criticisms of contemporary materialism and situate itself in a philosophical space which is not 'pre-' but 'post-' Nietzschean.

We can distinguish three key conceptions of transcendence. The first is that which was employed by the Ancients to describe the *cosmos*. Fundamentally, of course, Greek thought is a philosophy of immanence in that the perfect order of things is not an ideal, a model which is located elsewhere than in the universe, but is on the contrary wholly incarnated within its fabric. As you will recall, the divine principle of the Stoics, as distinct from the God of the Christians, is not a Being external to the world, but is so to speak its very organising principle. However, as I have already indicated in passing, we can also say that the harmonious order of the cosmos is nonetheless transcendent in relation to humans, in the specific sense that they have neither created it nor invented it. On the contrary, they discover it as a reality that is external to and superior to them. The word 'transcendent' here means in relation to humanity, designating a reality which exceeds us, without however being located elsewhere than in the universe. A transcendence on earth rather than in heaven.

A second conception of transcendence, quite different and even opposed to the first, applies to the God of the major monotheistic religions. It refers simply to the fact that the supreme Being is – contrary to Greek divinity – 'beyond' the world which He made, both external to and superior to creation as a whole. Contrary to the Stoic conception of divinity, which merges with the harmony of the natural order, and is consequently not located outside of it, the God of the Jews, Christians and Muslims is entirely 'supernatural'.

Here is a transcendence not merely relative to humanity (like that of the Greeks), but also relative to the universe itself, conceived entirely as a creation whose existence depends upon a Being exterior to it.

A third form of transcendence, different from the two versions described above, can also be formulated. It already takes root in the thought of Kant, and follows its course through the phenomenology of Husserl. It can be summarised in Husserl's 'transcendence within immanence'. The formula may not be elegant, but within it is concealed an idea of great profundity. Here is how Husserl himself apparently preferred to describe it to his students – for, like many of the great philosophers, Husserl was first and foremost a remarkable teacher. He would take a cube or rhomboid – a box of matches, for example – and hold it up before his students, making them observe the following: that however we try to arrange the cube in question, one can never see more than three sides at one time, although we know there are in fact six sides. You may well reply, 'So? Why does this matter, and what can it contribute to philosophical purposes?' To which there is an answer: that first and last there is no such thing as omniscience, no absolute knowledge, since everything that is visible (the visible as symbolised by the three exposed faces of the cube) rests on a foundation of invisibility (the three hidden faces of the cube). All presence supposes an absence; all immanence supposes a hidden transcendence; all visible sides of an object suppose a side that is invisible.

Clearly this example is no more than metaphorical. What it signifies is that this transcendence is not a new 'idol', an invention of metaphysics or belief, nor is it make-believe about a world beyond, the purpose of which is to belittle the real in the name of the ideal. Rather, transcendence is a fact, a deduction, an undeniable dimension of human existence, inscribed at the heart of our common reality. And it is in this respect that transcendence – or, more specifically, the notion of transcendence here-and-now – cannot be merely trodden underfoot by the classic critique of idols as formulated by materialists and other adherents of deconstruction. In this sense, it is not metaphysical, and it is post-Nietzschean.

To define more precisely this new idea of transcendence – before moving on to some examples – a fruitful approach is to reflect, Husserl suggested, on the notion of a *horizon*. When you open your eyes to the world, objects always offer themselves against a background, and this background, as you proceed further into the world which surrounds us, continually displaces itself rather as the horizon does for a sailor, without ever resolving itself into a final and impassable background. Thus, from background to background, from horizon to horizon, you can never succeed in grasping on to anything which you can hold as a final entity, a supreme Being or a first cause which might guarantee the real in which we are immersed. And it is in this respect, precisely, that there is transcendence, something which always escapes us at the very core of what we are given, of what we see and touch – at the core of immanence itself.

Like the cube, whose several faces I can never see simultaneously, the reality of the world is never presented to me as transparency, as mastery. In other words, if we confine ourselves to the point of view of human finiteness, to the idea – as expressed by Husserl, again – that 'all consciousness is a consciousness of something', that all consciousness is therefore limited by a world external to itself and consequently, in this sense, *finite*; then we must correspondingly admit that human knowledge can never attain to omniscience, can never coincide with the point of view which Christians accord to God.

It is therefore by its refusal of closure, by its rejection of all forms of 'absolute knowledge', that this third type of transcendence does indeed seem to be a 'transcendence within immanence', willing only to confer rigorous meaning to human experience as formulated by a humanism freed from the illusions of metaphysics. It is truly 'within me', in my thought or in my sensibility, that the transcendence of values manifests itself. Although they are situated within me ('immanent'), everything unfolds nonetheless as if values impose themselves ('transcendent') upon my subjectivity from without – as though they come from elsewhere.

Consider for a moment the four great settings in which the fundamental values of human existence are played out: truth, beauty, justice and love. All four of which, whatever the materialists say, remain fundamentally transcendent for the particular individual, for you and me, as for everyone else. Let us simplify a little more: I cannot invent mathematical truths, nor the beauty of a work of art, nor the imperatives of the moral life; and when I 'fall' in love, as the phrase so accurately describes it, I cannot choose deliberately to do this. The transcendence of values is in this sense patently real. But it is also housed in concrete experience, not in a metaphysical fiction, nor in the form of an idol such as 'God' or 'Paradise' or 'Republic', or 'Socialism'. We can construct a 'phenomenology' of this experience, a simple description which starts from the sense of an inescapable necessity, from the awareness of not being capable of thinking or feeling differently on a particular subject: I can do nothing about it, 2 + 2 = 4, and this is not a matter of taste or subjective choice. The necessities of which I speak impose themselves on me as if they come from elsewhere, and yet, it is inside myself that this transcendence is present, and palpably so.

In the same way, the beauty of a landscape or a piece of music imposes itself, 'bowls me over', transports me, irrespective of choice. And in the same way, I am not at all persuaded by the argument that I merely *choose* ethical values, that I decide for example to be anti-racist: the truth is rather that I cannot think otherwise, that the idea of a common humanity asserts itself with its attendant baggage of notions of justice and injustice.

There exists, well and truly, a transcendence of values, and this is the proposition embraced by a humanism without metaphysics (as opposed to a materialism which claims to explain everything by reducing everything, and without ever succeeding). Not embraced out of impotence, but in full awareness, because the experience of it is sovereign, and no materialism can arrive at an adequate explanation of it.

But if there is transcendence, why must it be 'within immanence'? Quite simply, because from this point of view, values are no longer imposed upon us in the name of authority nor deduced from this or that metaphysical or theological fiction. It is true that I discover – rather than invent – the truth of a mathematical proposition, the magnificence of the ocean or the legitimacy of the rights of man, but nonetheless it is unquestionably within me that these things are discovered, and nowhere else. There is no longer a heaven of metaphysical ideas, no God – or at least I am not obliged to think so in order to accept the idea that I am in the presence of values that are at once beyond me, yet nowhere to be found except within me, manifest only inside my consciousness and conscience.

Let us take another example. When I 'fall' in love, there is no doubt that, unless my name is Narcissus, I am wholly in thrall to a being exterior to myself, an other who eludes or is distinct from me, and whom I come to depend upon. In this case, too, there is transcendence. But it is also clear that it is in myself that I sense this transcendent reality of the other. It is located, so to speak, in that part of my person that is most intimate and private, in the sphere of feelings or, as we say, in 'the heart'. One could not find a more beautiful metaphor for immanence than this image of the heart. For the latter is at once the place of transcendence – of love for another as something irreducible to myself – but also of the immanence of love as the emotion most inward to myself. So, transcendence within immanence.

Materialism would reduce my experience of transcendence to the material realities which supposedly underlie it; whereas a humanism which has shed the naïve baggage still carried by modern philosophy (as described earlier), can offer instead a practical description without preconceptions: a 'phenomenology' of transcendence, as something settled at the core of my sense of self.

This is therefore why humanist *theoria* proves itself to be above all a theory of knowledge centred upon self-knowledge or, to use the language of contemporary philosophy, upon 'auto-reflection'. Contrary to materialism, which as I have suggested never succeeds in thinking its own thought, contemporary humanism sets itself to reflect upon the meaning of its own assertions, to become fully aware of them, to criticise and evaluate its own propositions. The critical spirit which already characterised modern philosophy from Descartes onwards must take a further step: instead of describing others, it will finally and systematically set about describing itself.

Theoria as 'Auto-reflection'

Here, once again, we can distinguish three ages of knowledge. The first corresponds to Greek *theoria*. Contemplation of a divinely ordained world, the endeavour to comprehend the structure of the *cosmos* – this was hardly, as we have seen, a knowledge indifferent to the question of values. Or, to use the terminology of the great twentieth-century German sociologist, Max Weber, it was not 'axiologically neutral' – meaning 'objective', disinterested or devoid of bias. As we have seen in the case of Stoicism, knowledge and values are inextricably linked, in the sense that the discovery of the cosmic nature of the universe implies certain moral purposes for human existence.

The second age appears with the modern scientific revolution, which sees emerge, in opposition to the Greek world, the idea of a knowledge that is radically indifferent to the question of values. In the eyes of the Moderns, not only does nature not give us any ethical pointers whatsoever, but no longer provides a model for us to imitate; furthermore, true science must be rigorously neutral in respect of values, on pain of being accused of partisanship and lack of objectivity. In other words: science must describe *what is*, not *what ought to be*, not what we should or should not be doing morally. As we say in philosophical jargon, science does not possess any *normative* (as opposed to *descriptive*) purpose. The biologist, for example, can demonstrate to you that smoking is harmful to your health, and on this point is entirely correct. On the other hand, whether from a moral point of view the act of smoking is or is not a fault, and, consequently, whether stopping smoking is an ethical obligation, he has nothing to say. It is for us to decide, on the basis of values which are not, as such, scientific. In this context, generally designated by the term 'positivist' and which came to predominate during the eighteenth and nineteenth centuries, science questioned itself rather less than it focused upon understanding the world as it is.

We can go further: the scientific method could not remain content with evaluating phenomena. There must come a day when, if only by obeying its own principles, it includes itself in the story. The critical spirit must arrive sooner or later at self-criticism, which is what modern philosophy is only beginning to come to terms with, but which Nietzsche and the materialists paradoxically refused to do. The genealogist and the deconstructionist worked wonders at firing bullets into metaphysics and religion, at breaking up our idols with a hammer, but on their own account it was a case of nothing doing. Their aversion to self-reflection constitutes their manner of seeing the world. Their clarity of analysis in respect of others is admirable, but it is equalled only by their blindness to their own positions.

A third age of knowledge arrives therefore to challenge but also to complete its precursor; namely an age of self-criticism or auto-reflection, which best defines contemporary post-Nietzschean humanism. This only began to occur just after the Second World War, when questions began to be asked about the potential misdeeds of a science in some sense responsible for the atrocities of Hiroshima and Nagasaki. This self-criticism was to continue more generally in all areas where the consequences of science might have moral or political implications – and latterly in the field of ecology or bio-ethics.

One could say that in the second half of the twentieth century science ceased to be essentially dogmatic and authoritarian, and began to apply to itself its own principles, method and critical spirit – which as a result became far more self-critical or 'auto-reflective'. Physicists questioned themselves about the potential dangers of the atom, or the possible ravages of the greenhouse effect; biologists asked themselves if genetically modified organisms present a risk for humanity, or if the technology of cloning is morally legitimate, and many other questions of a similar nature, which displayed a complete reversal of the nineteenth-century perspective. Science, no longer imperious and certain, learnt to challenge itself, slowly but surely.

From where proceeds the formidable expansion of the sciences of the past – history and historiography, historical geography, archivism – over the course of the twentieth century. History itself becomes the queen of 'human sciences', and here too it is useful to reflect for a moment upon the significance of the extraordinary expansion of these practices. Borrowing from the model of psychoanalysis, history promises us that by progressively reclaiming and mastering our past, by practising this collective and heightened form of auto-reflection, we shall come to a better understanding of our present and orient ourselves more effectively towards our future.

The historical sciences in the broadest sense, therefore, including a large swathe of the social sciences, take root progressively, and more or less consciously, in the conviction that history weighs upon our lives and destinies in proportion to our ignorance of the past. To know one's history, is, as in psychoanalysis, to work towards one's own emancipation, and a democratic ideal of liberty of thought cannot dispense with the study of history, if it is to approach the present without prejudices.

From which also proceeds the current and widespread error that philosophy devotes itself entirely to self-appraisal. There is some truth in this error: in effect, modern *theoria* has well and truly entered the age of auto-reflection. What is false, however, is the deduction that philosophy as a whole should remain fixed at this point – as if henceforth *theoria* was the one and only dimension of philosophy, as if the problem of salvation, notably, must be abandoned. I shall show in a moment that this is not so, that it remains more than ever alive, provided that we accept the need to formulate it in terms which are not those of the past. But first let us see how, in the perspective of a humanism without metaphysical claims, modern morality is also enriched with new dimensions.

The Deification of the Human

Nietzsche perfectly understood, even if he was to draw hostile conclusions and take an 'immoralist' direction, that the problem of morality arises from the moment that a human being posits sacrificial values, values 'superior to life'. There is a morality in play therefore as soon as principles present themselves to us, rightly or wrongly as so elevated, so 'sacred', as to seem worth risking or even sacrificing our lives for them.

I am sure, for example, that if you witnessed the lynching of somebody because of the colour of his skin, or on account of his religion, you would do what was in your power to help him, even if to do so was dangerous. And if you were to lack the courage, which is something everyone can understand, you would nevertheless admit to yourself that, morally, this is what ought to happen. And if the person being attacked was someone you love, then you would probably take enormous risks to save him or her. I give this small example – perhaps not a very likely scenario for us today, but all too likely in countries at war a mere plane journey away – in order to make the following reflection: counter to the inevitable logic of a thoroughgoing materialism, we continue to believe (whether or not we profess to be materialists) that certain values could, in a given situation, lead us to risk our lives.

In the early 1980s, when Soviet totalitarianism was still very much in place, German pacifists adopted a detestable slogan, *Lieber rot als tod* ('Better red than dead') – in other words, better to submit to oppression than risk death by resisting it. Evidently the slogan did not convince everyone, and there are still many people – not necessarily 'believers', either – who believe that the preservation of one's own life, infinitely precious though it may be, is not necessarily and in all circumstances the only value that counts. I am even convinced that, if need be, my fellow citizens would still take up arms to defend their neighbours or resist a totalitarian menace – or at the least, that such an attitude, even if they did not themselves have the courage to carry it through, would not strike them as either contemptible or ridiculous.

Sacrifice, which returns us to the notion of a value regarded as *sacred* (both from Latin, 'sacer'), paradoxically retains, even for the committed materialist, an aspect which can almost be described as religious. It implies, in effect, that we admit, however covertly, the existence of transcendent values, superior to our material and biological existence.

It is simply the case – and it is here that I would wish finally to identify what might be new about a humanist ethics in a contemporary context, as distinct from the morality of the Moderns – that the former motives for sacrifice have long departed. In our Occidental democracies, at least, very few individuals indeed would still be willing to sacrifice their lives for the glory of God, or for the homeland, or for the revolutionary proletariat. On the other hand, their freedom, and – still more likely – the lives of those they love, might well strike them in certain extreme circumstances, as worth fighting for. In other words, the radical immanence so dear to materialism (requiring a renunciation of the sacred, along with the very notion of sacrifice) has not in any way replaced the versions of transcendence formerly on offer – whether God, homeland or revolution. Instead, new forms of transcendence have intervened, 'horizontal' rather than vertical: rooted in our humanity, in other beings who are in the same frame as ourselves, rather than vested in abstract entities located above our heads.

In this respect, it seems to me that the evolution of the contemporary world has involved the intersection of two broad tendencies. There has been a *humanising of the divine*. To give an example: one could argue that the universal declaration of the rights of man is no more than (and again Nietzsche saw this clearly) a 'secularised' Christianity, in other words a restatement of the content of the Christian religion without belief in God being a requisite. And there is no doubt that we are living through a reversal of divinisation, or a *making sacred of the human*, in the sense I have just defined: it is only on behalf of another human being that we are prepared, in the case of necessity, to undertake risks, and certainly not to defend the abstract entities of the past. Because no one any longer believes that, in the words of the Cuban national anthem, 'to die for the homeland is to live for eternity'. Of course, we can remain patriotic, but the nation as such has changed meaning: it refers less to a territory than to its human inhabitants, and it is less a repository of nationalism than of humanism.

If you wish for an example, if not a proof, read the short but very important book by Henri Dunant entitled *A Memory of Solferino*. Dunant was the founder of the International Committee of the Red Cross, and thus the founder of modern humanitarianism, to which he dedicated his life. In this little book he describes how his extraordinary vocation came about. On a business assignment in 1859, reluctantly forced to cross the battlefield of Solferino (in its immediate aftermath), he witnessed a world of absolute horror. Thousands of dead and, worse still, countless wounded who were slowly dying in conditions of appalling suffering, without help or assistance of any kind. Dunant spent forty-eight hours, up to his elbows in blood, assisting the dying.

He drew an exemplary lesson from his experience, which is at the origin of a veritable moral revolution, that of modern humanitarianism and its protocols: according to which a soldier, once he is fallen, wounded and disarmed, ceases to belong to a particular nation or camp, but reverts to being a man, a simple human being who, as such, earns the right to be protected, assisted, cared for – irrespective of his participation in the conflict. Dunant here echoed the fundamental inspiration of the 1789 Declaration of the Rights of Man: every human being merits respect without regard for community or for ethnic, linguistic, cultural or religious allegiances.

But Dunant goes further than this, in that he asks us equally to disregard national allegiances, to the extent that the humanitarian mission, in this respect heir to Christianity, asks us to treat our enemy, in so far as he is reduced to a state of harmless humanity, in the same way as we would treat a friend.

As you see, we are a long way from Nietzsche, whose aversion to the notion even of compassion led him to detest all forms of charitable action, under suspicion of perpetuating Christianity, to the point of his literally jumping for joy when he learnt that an earthquake had struck Nice or a cyclone had devastated Fiji. Nietzsche strayed, of this there can be little doubt, but his diagnosis is not entirely wrong: by wearing a human face, the sacred does not diminish: the transcendent lives on, even lodged in the immanent, in the heart of man. Instead of deploring the situation, with Nietzsche, it is precisely this shift which must be thought afresh, if we are to stop living in denial together with the materialist: who recognises in his private experience the existence of values which bind *absolutely*, yet commits himself on a theoretical level to defending moral *relativism*.

It is on this basis that we can now raise our sights to a consideration of salvation, or at least to a consideration of what takes its place.

Rethinking Salvation

I would like to end by proposing three topics for reflection, on the manner in which a humanism without metaphysics might today give new life to the ancient problem of wisdom. First, the requirement of an 'enlarged thought'; second, the wisdom of love; third, the experience of mourning.

The Kantian notion of an 'enlarged thought', which I mentioned at the end of the chapter on modern philosophy, takes on a new significance after Nietszche. It no longer merely designates, as for Kant, the need for a critical spirit, a disposition to see the other side of a question ('putting oneself in the place of others, so as to understand their point of view'), but well and truly a new way of responding to the question of life's meaning. I would like to say a word about this, before indicating some of its points of connection with the question of human salvation.

In contrast with a 'restricted' vision, the horizon of an enlarged thought could be defined, initially, as one that manages to displace itself, 'to put itself in the place of another' – the better to understand, but also to try, in a movement of return upon itself, to look upon its own judgements as if they were those of another. It is this aspect that requires the auto-reflection of which we spoke earlier: to become properly conscious, one must situate oneself in some sense at a distance. Whereas the restricted self remains bogged down in its place of origin, to the point of believing that this is the only possible community, or at least the only proper and legitimate community, the enlarged spirit manages, by occupying in so far as possible the point of view of the other, to contemplate the world in the guise of a benevolent and disinterested spectator. Agreeing to unseat his initial and inherited way of seeing and to remove himself from the closed circle of egocentrism, he is able to penetrate customs and values remote from his own; then, returning to himself, he can be aware of himself in a distanced and less dogmatic fashion, and can thereby enrich his own view of things.

In this respect, and it indicates how deep are the intellectual roots of humanism, the notion of 'an enlarged thought' is continuous with that of human 'perfectibility', which Rousseau isolated as the specifically human, as opposed to animal, property. Both notions suppose the idea of an extended liberty of action, considered as the faculty of self-withdrawal from a particular condition in order to accede to an increased universality, whether individual or collective – that of education on the one hand, or culture and politics on the other – in the course of which is effected what one might call the humanising of the human. It is precisely this humanising process which gives all its meaning to life and which, in the quasi-theological meaning of the term, 'justifies' a life.

In my book *What is a Good Life?* I quoted at length from a speech given by the great Anglo-Indian writer V.S. Naipaul, on the occasion of his acceptance of the Nobel Prize for Literature in 2001. The passage in question seems to me to describe to perfection this experience of an 'enlarged thought' and the benefits it brings, not only in the writing of a book but also more profoundly in the conduct of a life. In the speech, entitled 'Two Worlds', Naipaul recalls his childhood in Trinidad and evokes the limitations inherent to the life of these small communities, enclosed upon themselves and folded back upon their particularisms:

> We Indians, immigrants from India . . . lived for the most part ritualised lives, and were not yet capable of self-assessment, which is where learning begins . . . In Trinidad, where as new arrivals we were a disadvantaged community, that excluding idea was a kind of protection; it enabled us – for the time being, and only for the time being – to live in our own way and according to our own rules, to live in our own fading India. It made for an extraordinary self-centredness. We looked inwards; we lived out our days; the world outside existed in a kind of darkness; we inquired about nothing. (Extract from V.S. Naipaul's Nobel Prize acceptance speech)

Naipaul goes on to explain how, once he became a writer, these 'zones of shadow' which surrounded the growing child – that is, everything that was more or less there and present on the island, but which self-absorption prevented him from seeing: the native population, the New World, India, the Muslim universe, Africa, England – became subjects of preoccupation which enabled him to establish a distance, and one day write a book about the island of his birth. As you can see, his entire itinerary as a man and a writer – the two are strictly inseparable here – has consisted of enlarging his horizon by making a profound effort of 'decentering,' uprooting himself with a view to being able to penetrate the 'zones of shadow' in question.

Then he adds this, which is perhaps the essential:

> The distance between the writer and his material grew with the two later books; the vision was wider. And then intuition led me to a large book about our family life. During this book my writing ambition grew. But when it was over I felt I had done all that I could do with my island material. No matter how

> much I meditated on it, no further fiction would come. Accident, then, rescued me. I became a traveller. I travelled in the Caribbean region and understood much more about the colonial set-up of which I had been part. I went to India, my ancestral land, for a year; it was a journey that broke my life in two. The books that I wrote about these two journeys took me to new realms of emotion, gave me a world-view I had never had, extended me technically. (Extract from V.S. Naipaul's Nobel Prize acceptance speech)

No renunciation, here, nor any disowning of the particularities of his origin. Merely a distancing, an extending (it is striking that Naipaul himself uses a vocabulary of enlargement) which enables him to grasp these particularities from another perspective, less immersed, less egocentric – by means of which his writing, far from standing still, like the local craft industry, managed to elevate itself to the level of 'world literature'. By which I mean it is not reserved for the 'indigenous' population of Trinidad, nor even a former colonial readership, because the itinerary it describes is not exclusively particular: it possesses a universal human meaning, which, beyond the particularity of Naipaul's circumstances, is able to affect all readers.

The literary and even existential ideal traced by Naipaul in these pages requires that we uproot ourselves from our egocentrism. We need others and otherness in order to understand ourselves; we need their liberty, even their happiness if possible, to accomplish our own lives. In this sense, the consideration of morality points towards a deeper question of meaning.

In the Bible, to know means to love: traditionally, to speak of knowing somebody 'biblically' means to have carnal knowledge. The question of meaning is a secular version of this biblical equation: if knowing and loving are one and the same, then what must give sense above all to our lives – at once an orientation and a meaning – is indeed the ideal of an enlarged horizon. This alone, by its *invitation au voyage*, its exhorting us to come out of ourselves the better to find ourselves – which is Hegel's dialectical definition of 'experience' – enables us better to know and love others.

This is perhaps the one and only answer to the question: 'What is the point of growing up?' To enlarge our vision, to learn to relish the singularity of others, and on occasion – when this love of the other attains its greatest intensity – to experience the abolition of time. In which we succeed, if only momentarily (as the Greeks exhorted us) to free ourselves from the tyranny of past and future, to inhabit a present which is finally serene and cleansed of guilt. It is at this point– where the fear of death has no reality – that the question of meaning intersects with that of salvation.

But first I would like to press further the question of whether there exists a 'wisdom of love', a vision which allows us to understand fully the reasons why it alone, in a humanist perspective at least, gives meaning to our lives.

I would like to begin with a very simplified analysis of what constitutes a work of art. In whatever realm, the work of art is always initially defined by the particularity of its cultural context of origin. It is always historically and geographically marked by the epoch and the 'spirit' of the people among whom it originates. This might be described as its 'folkloric' aspect (from German *Volk*, meaning 'people'), its debt to a common vocabulary or vernacular, if you like. We can tell immediately, without being in any way specialists, that a canvas by Vermeer belongs neither to the Asiatic world nor to the Arab world; that it obviously cannot be situated in terms of contemporary art, but rather belongs with Northern European art of the seventeenth century. Similarly, a few bars are often enough to tell us that a piece of music is Eastern or Western, that it is classical or modern, that it is religious music or dance music, and so on. Besides, even the greatest works of classical music borrow elements from popular dances, whose national characteristics are never far from the surface. A polonaise by Chopin, a Hungarian rhapsody by Brahms, the Romanian Dances of Bartok make these connections explicit. And even if it is not overt, the particularities of origin always leave their trace, and no matter how great a work of art, how universal its appeal, it never entirely breaks with its links to a place and a date.

However, it is equally the case that a great work is distinguished from a folkloric artefact in not being tethered to a particular 'people'. It raises itself to the 'universal'; it addresses itself potentially to the whole of humanity. This is what Goethe referred to, in terms of literary exchanges and relations, as *Weltliteratur* (world literature) – with which the notion of 'globalisation' has less than nothing in common: the access of the work of art to world status is not obtained by flouting its particularities of origin, but by assuming these from the outset, as its nourishment, so as to transfigure them in the space of art and make of them something other than simple folklore.

As a result, the particularities of origin are integrated into a larger context, to form an experience large enough to be potentially common to all of humanity. Which is why the work of art possesses the distinction of speaking to everyone, whatever the time or place in which we live.

Let us take a step further. In an attempt to understand Naipaul, I make use of two key concepts: particular and universal. The particular, to be specific, resides in the experience described by the writer, his point of departure: the small island and, more precisely, at the heart of the island, the Indian community to which Naipaul belonged. And the writing does absolutely concern a particular reality, with its own language, its religious traditions, its cuisine, its rituals etc. And then, at the other end of the spectrum, if you like, there is the universal. By which is meant not merely the vast world of others, but also the purpose of the itinerary to which Naipaul commits himself when he takes on the 'zones of shadow', or those elements of otherness which on first acquaintance he neither knows or understands.

What I would like you to understand, since it is crucial to grasp the senses in which love imparts meaning, is that between these two realities – the particular with its focus, and the universal which potentially includes all of humanity, there is room for a middle term: the singular, or the individual. And it is this latter reality, and this only, which is the object of our loves and the bearer of meaning.

Let us try to make sense of an idea which, quite simply, is the beam supporting the entire philosophical edifice of a secular humanism. To help us to see this more clearly, I will begin with a definition of singularity, inherited from German romanticism. If, as has been the case since classical antiquity, we designate by the term 'singularity' or 'individuality' a distinctive quality which is not merely that of the particular case, but graduates towards a broader horizon, to attain greater universality, it is immediately apparent that the work of art offers the most perfect model. And it is because they are, in this precise sense, the authors of singular works, at once rooted in their culture and epoch of origin, but at the same time capable of addressing themselves to all men in all ages, that we still read Plato or Homer, Molière or Shakespeare, and that we still listen to Bach or Chopin. The same is true of all great masterpieces and even great historical monuments: we can be English and Protestant and yet be profoundly moved by the temple of Angkor Wat, by the Great Mosque of Cairo, by a canvas of Vermeer or a scroll of Chinese calligraphy. Because they have raised themselves to the supreme level of 'singularity': meaning that they have dared to satisfy themselves neither with the particularities which formed them (as they form every individual) nor with an abstract, disembodied universality (like a chemical or mathematical formula, for example). The work of art worthy of the name is neither a local artefact nor is it a universal denuded of touch and taste, as is the product of pure scientific research. And it is to this singularity, this individuality that is neither entirely particular nor entirely universal, that we respond so powerfully.

From which you may also see how the notion of

singularity links directly to our ideal of an enlarged thought: by uprooting myself to become another, by enlarging the field of my experiences, I become singular – because I go beyond the particularities of my origins to accede, not to pure unmediated universality, but to a broader and richer awareness of the possibilities which are those of humanity as a whole. One simple example: when I settle in another country to learn a foreign language, I enlarge my horizon continually, whether I am aware of it or not. I afford myself the means of entering into communication with a larger number of people; and an entire culture is attached to the language I am discovering, so that I enrich myself incomparably by this new and external addition to my original particularity. In other words, singularity is not merely the primary characteristic of a work of art – this 'thing' that is external to me – but is also a subjective and personal attribute of the human individual as such. And it is this attribute, to the exclusion of all others, which is the primary object of our love for each other. We never love the particular as such, nor the universal in its abstraction and vacancy. Who would fall in love with a hedgehog or an algebraic formula?

If we hold on a little longer to this notion of singularity, to which the ideal of an enlarged horizon has led us, we must add the dimension of love: because it is love that gives its ultimate justification and meaning to the whole business of 'enlargement' which can and should guide human experience. Considered as such, it is the fulfilment of a humanist soteriology (the branch of theology that deals with salvation), the only plausible response to the question of life's meaning – in respect of which, once again, a humanism without metaphysics can easily look like a secularised Christianity.

We may be assisted in understanding this question by a fragment in Pascal's *Pensées*, where he quizzes himself, in effect, about the exact nature of the objects of our affection, and of the self that experiences affection:

> What then is this 'I'?
>
> Suppose a man places himself by a window to see those who pass by. If I pass by, can I say that he placed himself there to see me? No; for he does not think of me in particular. But does the man who loves someone on account of her beauty really love that person? No; for the smallpox, which will kill her beauty while sparing her person, will cause him to love her no more.
>
> And if someone loves me for my judgement, or for my memory, he does not love me, for I can lose these qualities without losing myself. Where, then, is this 'I', if it be neither in the body nor in the soul? And how love the body or the soul, except for those qualities which do not constitute an 'I', since they are perishable? For it is impossible and would be unjust to love the soul of a person in the abstract and for whatever qualities might be therein.
>
> We never, then, love a person, but only qualities.
>
> Let us therefore no longer jeer at those who are honoured on account of rank and office; for we only ever love a person on account of borrowed qualities. (*Pensées*, 323)

The conclusion usually drawn from this text runs as follows: the 'I', which Pascal constantly refers to as 'hateful', because it is always more or less vowed to egotism, is not a tenable object of love. Quite simply because we all tend to attach ourselves to particularities, to the 'external' qualities of those we claim to love: to their beauty, strength, humour, intelligence etc. This is what initially seduces us. But given that such attributes are bound to fail, love sooner or later gives way to weariness and boredom. And this, for Pascal, is our most common experience:

> He no longer loves the person whom he loved ten years ago. I can well believe it. She is no longer the same, nor is he. He was young, and she too; now she is quite different. He would perhaps love her still, were she as she was before. (*Pensées*, 123)

Yes, sadly. Far from having loved in the other person what we understood to be their most intimate essence – what I have been calling their singularity – we merely became attached to their particular and consequently entirely abstract qualities, which could as easily be found in any number of other people. Beauty, strength, intelligence etc. are not the preserve of this or that person, nor are they linked in any inward and essential sense to the 'substance' of this person as opposed to that person; these qualities are so to speak interchangeable. If Pascal's bored husband persists in his folly, he will probably divorce her in order to find a younger and more beautiful woman, just like the one he married ten years earlier.

Well before the German romantics of the early nineteenth century, Pascal discovered that the irreducibly particular and the interchangeably abstract and universal, far from being opposites, 'merge into each other', and are but two sides of the same coin. Reflect for a moment on the following all-too-common experience: you telephone a friend and you simply say, 'Hello, it's me', but this tells them nothing about who is speaking. This 'I' is abstract and lacking in singularity because everyone in the world calls themselves 'I'. Only by taking other information into account – your voice, for example – will your friend be able to identify you. But not by simple reference to an 'I' that remains paradoxically generalised and unlovably abstract.

In the same way, I think I have penetrated to the quick of another being, to what is most essential and irreplaceable about the beloved by loving them for their abstract and most undifferentiated qualities, but the reality is quite different: all I have identified are attributes as anonymous as a badge of office occupation or the letters that come after a name. In other words, the particular is not the same as the singular.

And we need to grasp that singularity alone, which transcends equally the particular and the universal, can be the proper object of love.

If we content ourselves with itemising particularities, we end up by failing ever to love anybody, in which case Pascal is correct: let us stop jeering at those who value only the borrowings of rank and office. After all, whether we go after beauty or medals comes down to the same thing: the first is (almost) as external to the person as the second. What makes an individual lovable, what creates the conviction that we could continue loving them even if their looks are ravaged by illness, is not reducible to an external attribute, a quality, however important it may be. What we love in the beloved (and are loved for in return, at least potentially), is neither pure particularity nor abstract universal attributes, but the singularity which distinguishes and renders he or she unlike any other. Of the one we love, we may say, affectionately, with Montaigne, 'because he was he, because I was I', but not, 'because he was handsome, strong, intelligent . . .'

And this singularity, you may be sure of it, was not handed out at birth. It is formed of a thousand details, and habits, of which moreover we are not even conscious. It is formed over the course of existence, and through experience; which is why, precisely, it is irreplaceable. Hedgehogs are all alike. As are kittens. Adorable, certainly. But it is only slowly, when it begins to smile, that the child becomes *humanly* lovable. From the moment he enters into a specifically human history, that of his relation to others.

In this sense, we can reinvest the Greek ideal of the 'eternal instant': this present moment which is freed from the anxieties of mortality – by virtue of its singularity, and because we regard it as irreplaceable, preferring to weigh it on its own terms rather than discard it in the name of nostalgia for what came before or hopes for what might follow.

It is here, once more, that the question of meaning connects with that of salvation. If withdrawal from the particular and exposure to what is universal are what create singularity of experience – if this double process gives singularity to our lives and allows us recognise what is singular in others – it offers us simultaneously the means of enlarging our thought and of acceding to moments of grace, where the fear of death (linked as it is to dimensions of time that are outside the present) is itself removed.

You might object that, compared to the doctrine of Christianity – whose promise of the resurrection of the body means that we shall be reunited with those we love after death – a humanism without metaphysics is small beer. I grant you that amongst the available doctrines of salvation, nothing can compete with Christianity – provided, that is, that you are a believer. If one is not a believer – and one cannot force oneself to believe, nor pretend to believe – then we must learn to think differently about the ultimate question posed by all doctrines of salvation, namely that of the death of a loved one.

There are, it seems to me, three ways of considering the loss of a loved one and three ways of preparing yourself for it. We can be tempted by the counsels of Buddhism, which can be reduced to a fundamental principle: do not become attached. Not from indifference – Buddhism, like Stoicism, speaks up for human compassion and the obligations of friendship, with the precaution that if we allow ourselves to be trapped by the net of attachments in which love invariably entangles us, we are without doubt preparing the worst of sufferings for ourselves: because life is a state of flux and impermanence, and human beings are perishable. We do not deprive ourselves only of happiness and serenity, in advance of the fact, but also of freedom. The words we use for these things are themselves suggestive: to be *attached* is to be *linked* or *bound*, as opposed to free; and if we wish to emancipate ourselves from the bonds forged by love, we must practise as early as possible that form of wisdom known as non-attachment.

Another response, diametrically opposed to the above,

characterises the great monotheisms – Christianity above all, since only Christianity professes the resurrection of the body as well as the soul. This consists of promising that as long as we practise love in God – in other words, a love that bears upon what is immortal in our loved ones rather than upon what is mortal – we shall experience the bliss of finding them again. In other words, attachment is not prohibited as long as it is correctly oriented. This promise is symbolised in the Gospels through the episode of the death of Lazarus, a friend of Christ. Christ weeps when he learns that his friend is dead – which Buddha would never allow himself to do. He weeps because, having taken human form, he is experiencing this separation as grief, as suffering. But he also knows, of course, that he will soon be united once more with Lazarus: that love is stronger than death.

Here are two forms of wisdom, then, two doctrines of salvation, which although opposed in almost every respect, deal nonetheless with the same problem: that of the death of loved ones. To put it bluntly, neither of these attitudes persuades me. Not only am I unable to prevent myself from forming attachments, I have no wish to do so. Nor am I in the least ignorant as to the sufferings to come – indeed I am already familiar with some of their bitterness. However, as the Dalai-Lama acknowledges, the only way of truly living according to the rules of non-attachment is to follow the monastic life, to be solitary (*monastikos*) in order to be free, to avoid all bonds. I believe that to be the case. Therefore I must renounce the wisdom of Buddhism, as I renounce that of Stoicism – with respect and esteem, but also with a sense of unbridgeable difference.

I find the Christian proposition infinitely more tempting – except for the fact that I do not believe in it. But were it to be true I would be certainly be a taker. I remember my friend, the atheist and historian François Furet, being asked on television what he would wish God to say to him were they ever to meet. To which he gave an immediate answer: 'Come quickly, your loved ones are waiting for you!' I would have given the same answer, and with the same undertow of disbelief.

What remains, then, other than to await the inevitable without paying it too much attention? Nothing, perhaps. Except, despite everything, to develop on one's own account, without any illusions, something resembling a 'wisdom of love' – as well as a love of wisdom. We each of us know that we must be reconciled with our parents before they die, whatever the tensions of that relationship. Because later, whatever Christianity may say, is too late. If we acknowledge that the dialogue with our loved ones must have a stop, then we must draw the consequences in this life.

It also strikes me that parents should not lie to their children about important things. I know several people who have discovered, after the death of their father, that he was not their biological parent – either that their mother had taken a lover, or that there was a concealed adoption. In every case, this kind of untruth causes considerable pain. Not merely because the belated discovery of the truth will always be unhappy; but above all because, after the death of the father who was not a father in the usual sense, it is impossible for the child turned adult to have that conversation: to grasp the meaning of a silence, or of a remark, or of an attitude which has left its trace on him, to which he would like to have given a meaning – but which he is now for ever prevented from doing.

To me this form of wisdom – a wisdom of love – is elaborated by each of us largely in silence. But I think that, to one side of Christianity or Buddhism, we can learn how to live and love as adults, even if this means thinking of death every day. Not out of morbidity, but to discover what needs doing, here and now, with those whom we love and whom we shall lose, unless they lose us first. I am convinced, even if I myself am still far from possessing it, that this type of wisdom exists, and that it is the crowning achievement of a humanism released finally from the illusions of metaphysics and religion.

IN CONCLUSION . . .

As you will have guessed, I love philosophy and above all I revere Kant's notion of an enlarged horizon of thought – upon which I have placed a lot of emphasis in these pages – as perhaps the central truth of modern philosophy and contemporary humanism. I think it permits us to create a *theoria* which gives the necessary space to self-reflection, an ethics which is open to the globalised world with which we are going to have to deal from now on, and also offers us a post-Nietzschean doctrine of salvation. Beyond these three great axes of enquiry, the dream of an enlarged thought also allows us to perceive differently – bypassing scepticism and dogmatism – the enigmatic prospect of there being a plurality of philosophical truths.

In general, the idea that there are several philosophical systems and that these do not agree with each other tends to provoke two responses: scepticism or dogmatism. Scepticism argues more or less as follows: since the dawn of time, different philosophies have done battle with each other without ever arriving at agreement about what constitutes the truth. And this plurality, because it is insurmountable, proves that philosophy is not an exact science; that it is lost in mist and unable to create a clearing for truth, which by definition must be single and unique. If there exist several versions of reality, and these fail to come to an agreement, we must then admit that none can claim seriously to hold the true answer to the questions we ask ourselves about knowledge, morality and salvation. Consequently, all philosophy is idle.

Dogmatism takes the opposite stance: certainly, there are several possible ways of looking at reality; and mine – the one I have ended up opting for – is manifestly superior to and truer than the others, which are nothing but a maze of endless errors. How many times must we listen to Spinozists telling us that Kant was off the wall, or Kantians denouncing the structural weaknesses of Spinoza?

Tired of these old debates, undermined by relativism, guilt-stricken too by the memory of its own imperialisms, the democratic impulse today sides willingly with positions of compromise, which, in the name of a commendable concern to 'respect differences', end up resigning themselves to slack notions of 'tolerance', 'dialogue', 'respect for others', to which it is not easy to assign a meaning.

The notion of an enlarged horizon suggests a different way forward. Sidestepping the choice between a pluralism of belief that is all façade, on the one hand, and a wholesale renunciation of convictions on the other, it invites us to extricate – case by case – what truth there might be in a vision of the world that is not ours, thereby affording us the means by which to understand it, and to take from it what we need for our own purposes.

I once wrote a book with a friend, André Comte-Sponville, the materialist philosopher whom I respect above all others. Everything stood between us: we are of the same age – room therefore for potential rivalry; politically he was coming from a communist background, and I from the republican right. Philosophically, he drew his inspiration entirely from Spinoza and the sages of the East, whereas mine derives from Kant and from Christianity. But instead of hating each other, we ended almost by trading places. By which I mean that, far from presuming that the other was acting in bad faith, we separately attempted to understand as fully as possible what might be persuasive and convincing about a vision of the world that is not ours.

Thanks to which, I have come to understand the grandeur of Stoicism, of Buddhism, of Spinozism – all those philosophies which invite us 'to hope a little less and love a little more'. I have understood, too, how the combined weight of past and future deadens our relish in the present; I have come to a greater liking for Nietzsche, even, and his doctrine of the innocence of becoming. As it happens I did not turn into a materialist, but I can no longer ignore materialism if I am to comprehend and describe certain aspects of human experience. In summary, I think that I enlarged the horizon that had been mine hitherto.

Every great philosophical system epitomises in the form of thought a fundamental human experience, just as every great work of art or literature translates human possibility into the most concrete and sensuous form. Respect for the Other does not after all exclude personal choice. On the contrary, it is its primary condition.

FURTHER READING

It would be easy to provide a bibliography, as we used to do in universities. The first hour of the philosophy course was spent in taking down a dictated list of a hundred and fifty titles, together with the secondary literature – all of which to be read without fail by the end of the year. The only problem is that this did not really serve any purpose; even less so today when you can find all the bibliographies you could wish for online within a few seconds. So I prefer to offer a short but 'reasoned' list for further reading, merely to call to your attention the few necessary books with which you should make a start – without trying to anticipate what will follow. And, to be quite honest, there is enough here to be getting on with.

Pierre Hadot, *What is Ancient Philosophy?* (Harvard University Press, 2002)

Jean-Jacques Rousseau, *Discourse on the Origin of Inequality* (translated by Patrick Coleman, Oxford University Press, 1999)

Immanuel Kant, *Groundwork of the Metaphysics of Morals* (translated by H. J. Paton, Harper Perennial Modern Thought, 2009)

Friedrich Nietzsche, *Twilight of the Idols* (translated by Duncan Large, Oxford University Press, 1998)

Jean-Paul Sartre, *Existentialism is a Humanism* (translated by Carol Macomber, Yale University Press, 2007)

André Comte-Sponville, *Le Bonheur, désespérément* (Editions Librio, 2003)

Martin Heidegger, 'What is Metaphysics?' and 'After Metaphysics', in *Basic Writings* (ed. David Farrell Krell, Harper Perennial Modern Thought, 2008)

INDEX

BOOKS BY LUC FERRY

THE WISDOM OF THE MYTHS

How Greek Mythology Can Change Your Life

Available in Paperback and eBook

Heroes, gods, and mortals. The Greek myths are the founding narratives of Western civilization: to understand them is to know the origins of philosophy, literature, art, science, law, and more. Indeed as Luc Ferry shows in this remarkable book, they remain a great store of wisdom, as relevant to our lives today as ever before. These classic stories provide the first sustained attempt to answer fundamental human questions concerning "the good life," the burden of mortality, and how to find one's place in the world—*The Wisdom of the Myths* will enlighten readers of all ages.

"Ferry's charm as a teacher bursts through on every page." —*Wall Street Journal*

A BRIEF HISTORY OF THOUGHT

A Philosophical Guide to Living

Available in Paperback and eBook

From the timeless wisdom of the ancient Greeks to Christianity, the Enlightenment, existentialism, and postmodernism, Luc Ferry's instant classic brilliantly and accessibly explains the enduring teachings of philosophy—including its profound relevance to modern daily life and its essential role in achieving happiness and living a meaningful life. This lively journey through the great thinkers will enlighten every reader, young and old.

"This superb primer proves that philosophy belongs at the center of life."
—*Publishers Weekly* (Starred Review)

"No dry academic, Ferry restores to philosophy a compelling urgency."
—*Booklist* (Starred Review)